INTELLIGENT COMMUNICATION SYSTEMS

Nobuyoshi Terashima
Graduate School of Global Information
and Telecommunication Studies
Waseda University
Tokyo, Japan

ACADEMIC PRESS
A Harcourt Science and Technology Company
San Diego San Francisco New York Boston
London Sydney Tokyo

This book is printed on acid-free paper. ∞

These materials were previously published in Japanese under the title of *The Intelligent Communication System: Toward Constructing Human Friendly Communication Environments.*

ACADEMIC PRESS
A division of Harcourt, Inc.
525 B Street, Suite 1900, San Diego, California 92101-4495, USA
http://www.academicpress.com

Academic Press
Harcourt Place, 32 Jamestown Road, London NW1 7BY, UK
http://www.academicpress.com

Library of Congress Catalog Card Number: 2001091273
International Standard Book Number: 0-12-685351-7

Printed in the United States of America
01 02 03 04 05 06 ML 9 8 7 6 5 4 3 2 1

CONTENTS

PREFACE

The information technology (IT) revolution is surely coming in this century, just as did the agricultural and industrial revolutions that have already so enriched our lives. As the IT revolution progresses, it is expected that almost all social structures and economic activities will be changed substantially.

In order for the IT revolution to penetrate our societies and enrich our lives, everyone in the world must have easy access to the information infrastructure and enjoy the use of any of the functions made available by that revolution. To accomplish this, the following basic functions have to be developed. Human-friendly human–machine interfaces should be provided to enable everyone, young or old, access to the information. Development tools have to be available for anyone to develop the new IT services. A more human-friendly communication environment is needed to allow people to communicate via the Internet as if they were gathered at the same place.

To fulfill these functions, the application of artificial intelligence (AI), such as natural language processing and knowledge engineering, to telecommunications will play an important role. The application of AI to telecommunication technology results in what is called the *intelligent communication system*. Research on the intelligent communication system includes the application of AI to telecommunications to produce human-friendly interfaces to telecommunication services, telecommunication description methods that are easy to use, and human-friendly telecommunication environments.

The intelligent communication system is a direct result of more than 10 years of industry experience, research activity, and education. In this book, the fundamentals of the foregoing research areas are described. For the research on telecommunication description methods, a description method based on state space is described. For the research on human-friendly interfaces for telecommunication services, AI applications that employ production systems, semantic networks, and predicate logic are described. For the research on human-friendly telecommunication environments, the concepts of *Telesensation* and *HyperReality* are described. Fundamental technologies such as computer vision are also discussed. Before launching into these research areas, the book first covers telecommunication fundamentals, telecommunication network structures, advances in telecommunication systems, information superhighways, and newly developed telecommunication systems.

In Chapter 1, IT, which is the convergence of information processing and telecommunication, is described. By combining information processing technology with telecommunications, more human-friendly communication interfaces are provided. Information technology provides not only telecommunication functions but also more human-friendly human–machine environments. Where we describe one of the IT architecture models, intelligent network (IN) architecture, the components needed for IN architecture are defined.

In Chapter 2, communication fundamentals, such as connection methods, the numbering plan, and protocols, are described. There are two connection methods: the connection type of communication and the connectionless type of communication. Communication by telephone is a connection type of communication. Communication by packet-switched network is a connectionless type of communication. In this chapter, the numbering plan of telephone service is described. By standardizing the numbering plan around the globe, someone in one country can telephone somebody in any other country.

In Chapter 3, communication network architecture is described. Initially, the telephone network was constructed. Then the computer network was built based on the telephone network according to advances in information processing technology. Recently, the Internet has been expanding throughout the globe. This chapter describes the network architecture of the telephone network, the network architecture of the computer network and the details of OSI protocol, and the network architecture of the Internet and the details of TCP/IP protocol.

In Chapter 4, the progress of telecommunication systems is described. Telecommunication networks have advanced greatly, from an analog network to a digital network. Initially the service-dependent networks were constructed for a data communication service and for a facsimile communication service. By integrating all of these networks via the digital network, the integrated services digital network (ISDN) was built.

In Chapter 5, several telecommunication systems, such as the data communication system, facsimile communication system, and videotex communication system, are described. With progress in telecommunication and information technology, various kinds of telecommunication services have been developed and put into practical use.

In Chapter 6, the information superhighways being developed in various countries are described. The idea of a national information infrastructure (NII) was proposed by the Clinton administration. After NII was proposed, many countries followed this initiative and devised their own concepts and development plans on information superhighways. Now under the umbrella of a global information infrastructure (GII), many countries are trying to build their own such highways.

In Chapter 7, newly developed telecommunication services are described. In this chapter, the newly developed telephone services, such as free phone service, source ID service, call forwarding service, and call waiting service, are described. Then mobile phone service is described. The number of mobile phone subscribers is increasing rapidly year by year. The potential applications of telecommunications, such as Continuous Acquisition and Lifelong Support (CALS) and electronic money, are described. The former provides the means, tools, or systems for conducting a business transaction at light speed. Electronic money and how to secure information transmitted over the network are focused on. The secure sockets layer and secure electronic transactions are described.

Chapter 8 describes the concept of the intelligent communication system, its system structure, its platform for a telecommunication system, and the knowledge base system that is a key component for constructing the intelligent communication system.

In Chapter 9, the design methodology for telecommunication services is described. AI theories, such as the state transition rule, graph theory, and predicate logic, are used for describing telecommunication services.

In Chapter 10, basic technologies of the intelligent communication system are described. Network components such as the terminal, computer, and network system are described by using the semantic network. Predicate logic is used for defining the syntax of dialog between human and computer. Symbolic logic is a basis of predicate logic. These theories are described here.

In Chapter 11, a next-generation communication environment, called *Telesensation*, is discussed. Through telesensation, an image, for example, of a scene from a natural environment or of a museum exhibit from a remote place is instantly transmitted through the communication links to viewers. Via stereoscopic display of such images using virtual reality (VR) technology, the viewers can enter the scene, a virtual world, and walk through it. Furthermore, the viewers can touch the leaves on a tree or the wall of the museum. They can behave as if they were actually present in that place. A further step, *HyperReality*, is introduced. In HyperReality, inhabitants, real or virtual, in reality their avatars, are brought together via the communication network and work or play together as if gathered in the same place. Several potential applications are also described.

Chapter 12 describes computer vision, a key technology for development of the intelligent communication system. Image analysis, image transformation, image recognition, and image synthesis are described, as is how to apply these technologies to the intelligent communication system.

Chapter 13 presents concluding remarks. Impacts on industry and society are described.

AUTHOR'S NOTE

This book is a direct result of over 10 years of research and education. My colleagues and I conceptualized a virtual-space teleconferencing system as a next-generation video conference system more than 10 years ago at ATR Communication Systems Research Laboratories, Kyoto. After that, I thought about a new concept that would provide a more human-friendly environment, as if we had been in a real world. In 1993 Professor John Tiffin of Victoria University of Wellington, New Zealand, visited ATR and examined the system. He was greatly impressed by its advances and tremendous possibilities. He had conducted distance education by interconnecting the main campus of Victoria University and a satellite campus at Taranaki. He was thinking about a more advanced distance education system. We talked about the possibility of applying the concept of a virtual-space teleconferencing system to distance education. After his visit to ATR, we started joint research on a next-generation distance education system. In 1994 I conceptualized *HyperReality* as a new paradigm for telecommunications. In 1996, I moved to Waseda University, Tokyo, as a full-time professor. I have focused on distance education as a potential application of HyperReality.

As a next-generation distance education system, John and I conceptualized *HyperClass*, by which a teacher and students, in reality their avatars, are brought together via the Internet to hold a class as well as to do cooperative work as if gathered in the same place. In 1998, a prototype system of HyperClass was developed. Using this system, we conducted the experiment on HyperClass by interconnecting Waseda University and Victoria University over the Internet. It was successful.

In December 2000, Queensland Open Learning Network, Australia, joined our project. We had a joint experiment on HyperClass by interconnecting three sites via the Internet. Our tasks were to handle a virtual Japanese artifact and to assemble the components into a computer. A Japanese teacher taught the history of Japanese artifacts and how to assemble components. Students of New Zealand and Australia learned by handling a virtual object by mouse and looking at it from various angles. This proved that it was very important not only to listen to the lecture but also to handle a virtual object directly. It was the epoch-making event for our project.

As mentioned in this book, the intelligent communication system provides an easy-to-use design method, such as the description method of telecommunications, the human-friendly interface to telecommunication users, and the human-friendly telecommunication environment. Through the experiment, HyperClass was proved to be useful for teacher and students. They can handle a virtual object in a human-friendly fashion. It is good not only for teaching but also for learning.

HyperClass is based on HyperReality. HyperReality is one of the key concepts of the intelligent communication system. The intelligent communication system provides a communication infrastructure for the development of communication services. The goal of telecommunications is to provide a human-friendly communication environment whereby human beings, real or virtual, at different locations are brought together via the communication network and talk or work as if gathered in the same real space.

Using the intelligent communication system, the communication system developers, the subscribers, and the communication service providers will receive the following benefits. Communication system developers can implement the communication system by means of the easy-to-use description methods and tools. Subscribers can interact with the communication system in a human-friendly fashion, for example, by using hand gestures or a natural language interface. Application service providers can, via the platform of HyperReality, make application programs easily. I hope this book will give readers insight into the information age and a hint at the conceptualization and development of the limitless applications in telecommunications

Finally, I would like to express my heartfelt thanks to Professor John Tiffin for his thoughtful suggestions to my work in establishing the concept of HyperReality and to Mr. Koji Matsukawa for his willing help to draw illustrations for the book. I also thank Ms. Anne Gooley of Queensland Open Learning Network, Australia, and Dr. Lalita Rajasingham of Victoria University, New Zealand, for their participation in the joint research on HyperClass.

1

INFORMATION TECHNOLOGY

In 1992 the International Conference on Global Survival was held in Stockholm, sponsored by the Institute for Future Studies of Sweden. The conference objective was to discuss global survival in the next millennium from the technical and social points of view. I was invited as a guest speaker to talk about information technology (IT) and its future prospects. I decided to talk about one of the potential fields of IT, a new concept named Telesensation.

I spoke about *telesensation*, a new concept that combines virtual reality (VR) with telecommunications, endowing telecommunications with realistic sensations. I coined the term to mean the integration of telecommunication and VR. Telesensation involves taking an image (for example, of a scene from a natural environment or a museum exhibit) gathered by camera from a remote place and transmitting that image over a communication network to viewers. Displaying the image on the screen stereoscopically by using VR technology, viewers can enter and walk through the virtual world. They can even touch the leaves on a tree or the wall of a museum. They can behave as if actually present in that place. Telesensation can break the bonds of time and space and contribute to reducing traffic on the road and is therefore environmentally friendly. The audience, clearly interested in the concept, posed many questions after my speech: When will it be put into practical use? What kinds of applications are developed based on the concept?

Figure 11.2 depicts a schematic of telesensation. A camera takes a picture of a street scene in Munich. The picture is then sent from Germany to Japan through a broadband integrated services digital network (ISDN). The picture is displayed stereoscopically by means of VR technology, and a viewer in Japan enters and walks through this virtual scene. He can go to the entrance of the building and walk inside. Or he can go behind the building and see what it looks like from there.

In 1996, the International Federation of Information Processing (IFIP) world congress '96 was held in Canberra, Australia, for which I was conference chair. The theme of the conference was IT—Global Horizon. The IT topics discussed included information processing, mobile communication, and teleteaching. In this context IT meant the combination of information processing and telecommunication.

Speaking at the closing ceremony, a historian from Australia referred to three epochs in human experience, spanning the past and the future. The first epoch was the agricultural revolution. Through the invention of agriculture, humans could produce foods. The second was the industrial revolution, by which engines and automatic machines were invented. The invention of powerful machines enabled the evolution of heavy industries such as the steel and power industries. The third epoch is the IT revolution, which will come in this millennium. Through the IT revolution, new industry will emerge. Electronic commerce on the Internet, manufacturing on demand, telecommuting, virtual school and virtual university, newspaper distribution via the Internet, and desktop publishing on the Internet will arrive soon. In this chapter technologies that will further push the IT frontier are discussed.

As stated before, IT is the integration of information processing and communication technologies. Automatic telecommunication technologies began with step-by-step switching systems, followed by crossbar switching systems and then by switching systems controlled by computers with stored memory. Information processing and data processing were enhanced with the invention of computers, and then the more advanced technologies, such as AI and knowledge engineering, were developed. Communication technology and information processing technology are also based on computers with stored memory. Thus advances in computer technology have advanced both information technology and communication technology. This has led to the integration of information processing technology and telecommunication technology—in other words, information technology.

1.1 INFORMATION TECHNOLOGY CONCEPT

With the invention of new telecommunication services, telecommunication networks for the services have been developed. The conventional telecommunication services, such as telephone and facsimile services, have been provided via the public telephone network. Video conferencing service has been provided by using the public network or dedicated lines. Data communication service has been provided by the public network or high-speed dedicated lines. Generally speaking, each service is provided by constructing a network suitable for the service. It takes a lot of money to construct, enhance, and maintain each of these networks. To overcome this problem, the integrated service digital network has been constructed to accommodate all of these services.

Recently the Internet has evolved, by which local area networks, long-distance lines, dedicated lines, and public analog/digital networks have been interconnected. Over the Internet, customers can easily access the network, send e-mail, access service providers such as Netscape Communicator and Internet Explorer, or access information providers. The number of customers on the Internet is increasing year by year. According to one forecast, the total number of users will reach 400 million by the end of 2002.

It will be very important to provide barrier-free and universal services to customers, young and old, around the globe. Users' requests are given in a variety of ways, such as spoken language, writing, gesture, and images. Somebody says in Japanese, "I would like to buy a book on IT, in particular on voice recognition." Or someone says in English, "I will go to Hawaii next week. Would you be kind enough to reserve two seats in business class on United Airlines." Or two people exchange e-mail messages over the Internet, one in English and the other in Japanese. Or someone handles a virtual object by hand gesture wearing a data glove in virtual space.

In the first example, spoken language is analyzed and converted into the canonical form of the sentence by a human–machine interface module. The system understands that the user would like to purchase a book on IT and then accesses the website of the bookstore and receives the answer "yes" or "no." This processing is done by an intelligent processing module. In the second example, the system analyzes the spoken language and understands the intention that the user would like to reserve two seats in business class on UA next week. This processing is done by a human–machine interface module. Then the system accesses the website of a travel agent and receives the answer. In the third example, the system analyzes the sentences by means of a human–machine interface module. The translation between Japanese and English is accomplished by an intelligent processing module. In the fourth example, the system analyzes a hand gesture and understands the meaning. This is done via a human–machine interface module. Then the system converts the gesture to the motion. According to the hand motion, the object is moved by an intelligent processing module.

As these examples show, human–machine interface modules and intelligent processing modules are needed to analyze, understand, and fulfill users' requests. To achieve this, these modules have to be installed in the system, which is running on the telecommunication network. The system comprises the communication network, terminals, human–machine interface modules, and intelligent processing modules, where human–machine interface modules are installed in the client stationed in the terminal, intelligent processing modules are installed in the server, and the client and server are interconnected over the communication network.

The structure of the IT system is shown in Figure 1.1. Its characteristics are as follows.

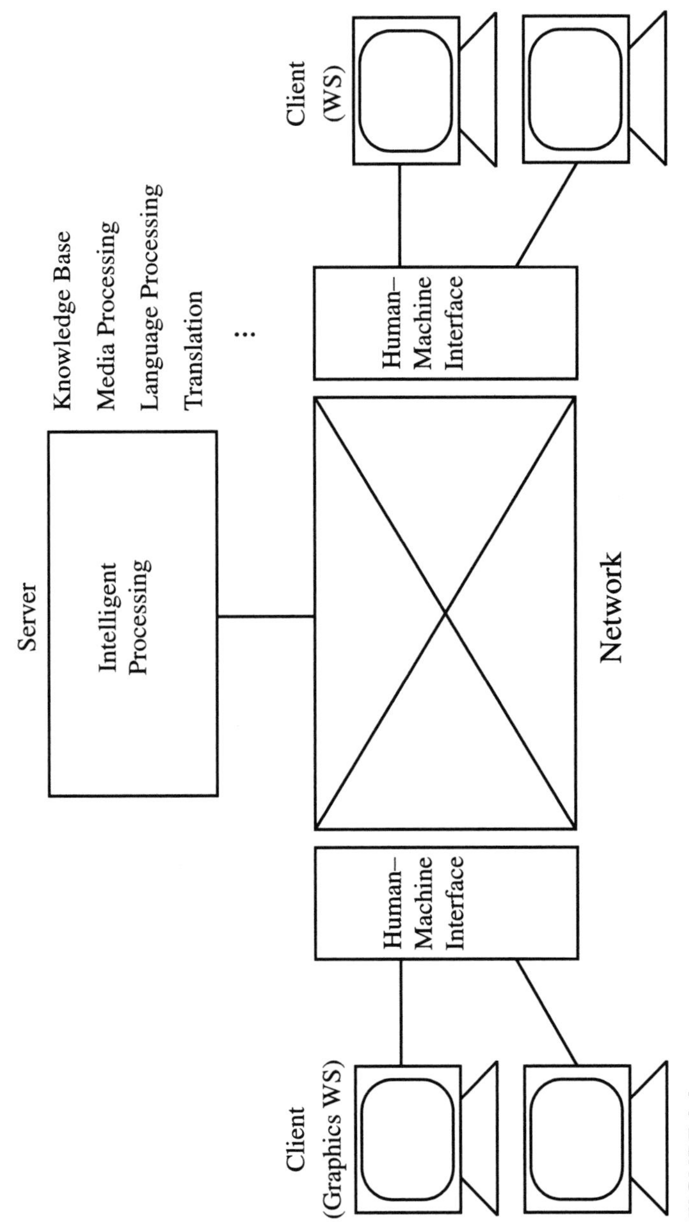

FIGURE 1.1 Schematic of the intelligent communication system.

(1) An IT system is composed of a communication network, terminals such as workstations and graphics workstations, human–machine interface modules, and intelligent processing modules.
(2) Users can access services through the terminals.
(3) The server has intelligent processing facilities, such as media conversion translation, or natural language processing facilities.
(4) The human–machine interface modules have natural language processing, speech processing, image processing, and gesture recognition facilities and provide human-friendly services to clients.

1.2 INTELLIGENT NETWORK CONCEPT

The next-generation communication network, called the *intelligent network* (IN) has been studied in many countries, especially the advanced countries (Figure 1.2). The functions needed for the IN are as follows.

(1) The network acts as a platform for information services. In concrete terms, connectivity between an information provider and a client must be fully available in the communication network. To achieve this, the network provides transmission paths that are transparent not only to information providers but also to clients with respect to the numbering plan, the fee policy, and the like.

(2) The network is independent of services and equipment. Many kinds of terminals and services will be installed in the network, so it should accommodate all kinds of services and equipment.

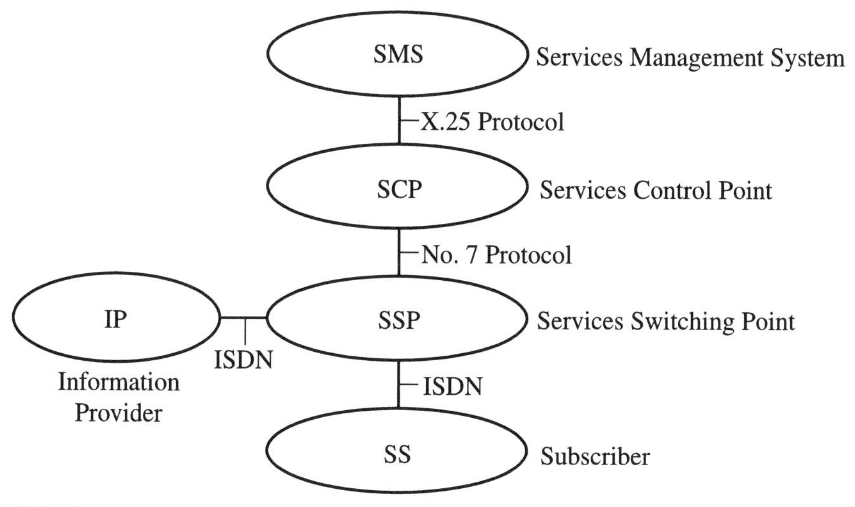

FIGURE 1.2 IN architecture.

(3) A network is connected to other networks, which are provided by the other different common carriers. Therefore, the interface for interconnection of networks has to be standardized.

As one IN architecture model, Bell Laboratories has invented the *advanced intelligent network* (AIN). In this architecture, the *functional component* (FC), a set of standardized call-control commands, has been introduced. The FC is service independent, so any services can use it for their implementations.

The service switching point (SSP) is a switching system that accommodates subscribers and information providers. It may be a stored-program-controlled switching system or an ATM switching system.

The services control point (SCP) includes the following modules.

- *Service logical program (SLP):* provides call processing functions
- *Service logical interpreter (SLI):* executes SLP according to the request for interconnection
- *Network interface database (NID):* stores the information concerning clients and networks
- *Network resource management (NRM):* manages the network resources for call processing

These modules may be installed in an SSP according to traffic conditions and may be transferred to an SSP that is located at the remote site. And SLI and NID may be used in any IN-based network.

The *service management system (SMS)* provides the functions of network operation, management, and maintenance. For example, the *service creation environment (SCE)* module supports the development of a new service. As the transmission protocols between a client and a network or between networks, the X.25, No. 7, and ISDN protocols are mainly used.

In many countries, especially advanced countries such as the United States, the European Union (EU), and Japan, new services, including computer telephony integration (CTI) services, have been developed on the intelligent network. At the same time, network architectural studies have been conducted. In the future, more advanced systems and services will be implemented and put into practical use based on IN architecture.

2

COMMUNICATION FUNDAMENTALS

2.1 CONNECTION-TYPE COMMUNICATION AND CONNECTIONLESS-TYPE COMMUNICATION

The objective of communication is the interchange of information between a source and its destination. One way to categorize telecommunication is into connection-type communication and connectionless-type communication. For *connection-type* communication, the source sends a message to its destination and receives acknowledgment from the destination. By comparison, for *connectionless-type* communication, a source sends a message to its destination without acknowledgment. The telephone is an example of connection-type communication. A letter or a postcard is an example of connectionless-type communication. Connection-type communication can be characterized by the fact that when the source gets no response from its destination, the source reissues the message until acknowledgment is returned. On the other hand, with connectionless-type communication, the source sends a message to its destination but expects no response from the destination.

To summarize, in connection-type communication, communication is complete when the source receives acknowledgment, which therefore takes time. In connectionless-type communication, on the other hand, the source only sends a message and expects no response, which therefore takes no time. This makes it more appropriate to use connection-type communication when the quality of the transmission line is not so good. However, when the quality is good, it is appropriate to use connectionless-type communication.

The Internet protocol is TCP/IP. Transmission Control Protocol (TCP) is a connection type of communication; Internet Protocol (IP) is a connectionless type of communication. When both TCP and IP have connection types we can transmit information from a source to its destination exactly, but it takes time. When we have high-quality transmission lines, it is sufficient to have TCP with connection type and IP with connectionless type.

Figure 2.1 shows an example of an exchange of information between a source and its destination. First, a request for connection is issued from the source to the destination. When acknowledgment is received from the destination,

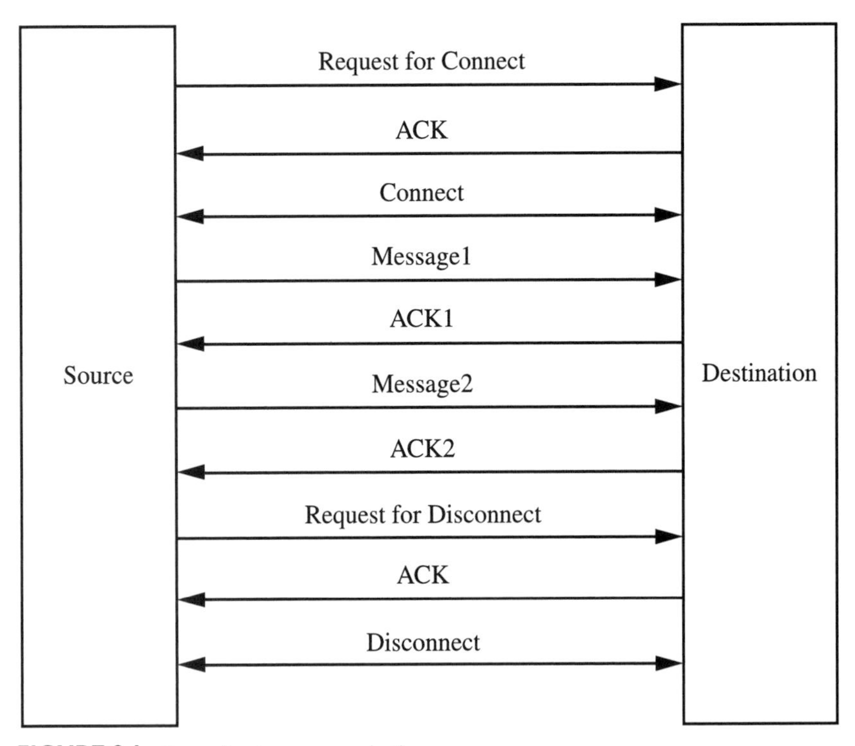

FIGURE 2.1 Connection-type communication.

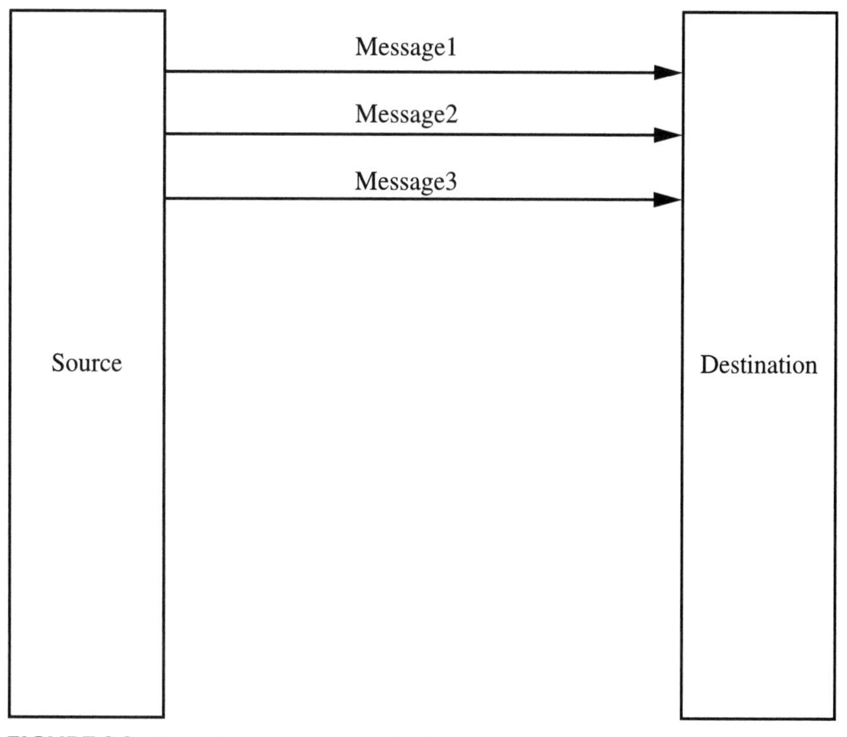

FIGURE 2.2 Connectionless-type communication.

the connection is established and then message 1 is issued. Also, acknowledgment 1 is received. This is followed by message 2 and acknowledgment 2. After the message has been sent, a request for disconnection is issued. When acknowledgment is received, the line becomes disconnected. An example of connectionless-type communication is shown in Figure 2.2, where messages 1, 2, and 3 are issued without acknowledgment.

For example, consider the making of a phone call. A source picks up a phone and dials the destination phone number. If the destination is idle, the destination phone rings and the source has a ringback tone. When the destination picks up the phone, the connection is established and the conversation starts. When the conversation finishes and either the source or the destination hangs up the phone, the link is disconnected. This is connection-type communication. Communication via telephone network, ISDN network, or the digital data switching network is connection-type communication.

On the other hand, with a *packet switched* network, a packet, which is composed of content and its destination address, can be transmitted to the destination

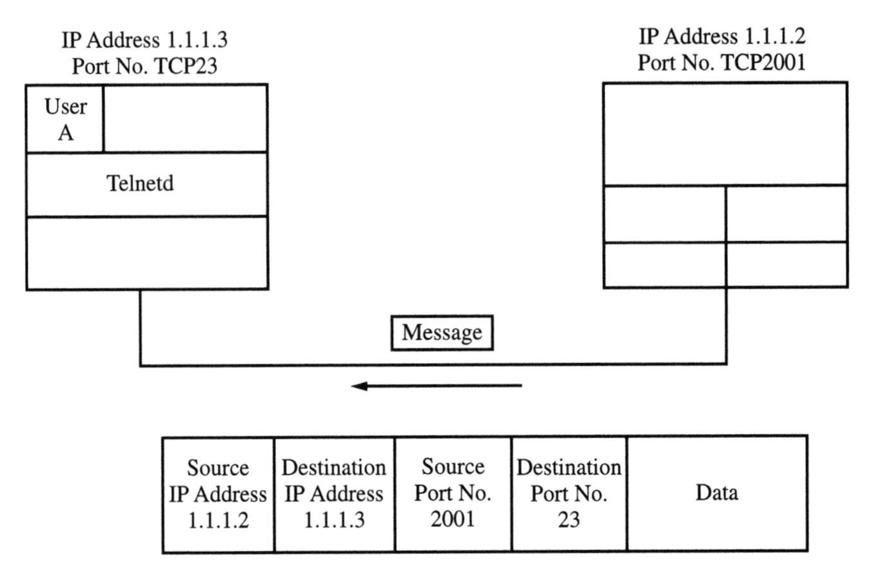

FIGURE 2.3 TCP/IP protocol process flow.

without establishment of the connection. The Internet and packet switched networks are classified as connectionless-type communications.

The process flow of the TCP/IP protocol is shown in Figure 2.3. Here, a message consists of a destination IP address, a source IP address, a destination port number, a source port number, and data to be transmitted. The message is sent from port number TCP2001 to port number TCP23.

2.2 NUMBERING PLAN

The objective of communication is to transmit a message from a source to its destination. When we write a letter we specify a destination. In the same way, we have to specify the destination address in the network message. The numbering plan allows this. In the case of telephones, we have such numbering plans as an international prefix, a country code, a toll number, a local number, and a subscriber number. For example, when a call is made to the United States from Japan, we dial a number such as 001-1-800-212-3141. The international prefix is the international carrier ID. Japan has such international prefixes as 001, 0041, and 0061.

2.3 PROTOCOL

The protocol specifies how to write a destination address and how to transmit a message over the network. A message is sent to the destination node via the neighboring nodes (see Figure 2.4). There are one or more processes at each node. The message is sent to the destination process. To transmit information to the destination process, the following action is required: First, information is transmitted to the neighboring node. This is performed under a data-link-level protocol. Second, information is transmitted to the destination node. This can be done under a network-level protocol. Third, information is sent to the destination process. It is done through a transport-level protocol. To transmit the information to the destination, it is necessary to establish the connection between the source process and its destination process and to disconnect when communication is over. Furthermore, it is necessary to specify the transmission method, such as duplex transmission (i.e.,

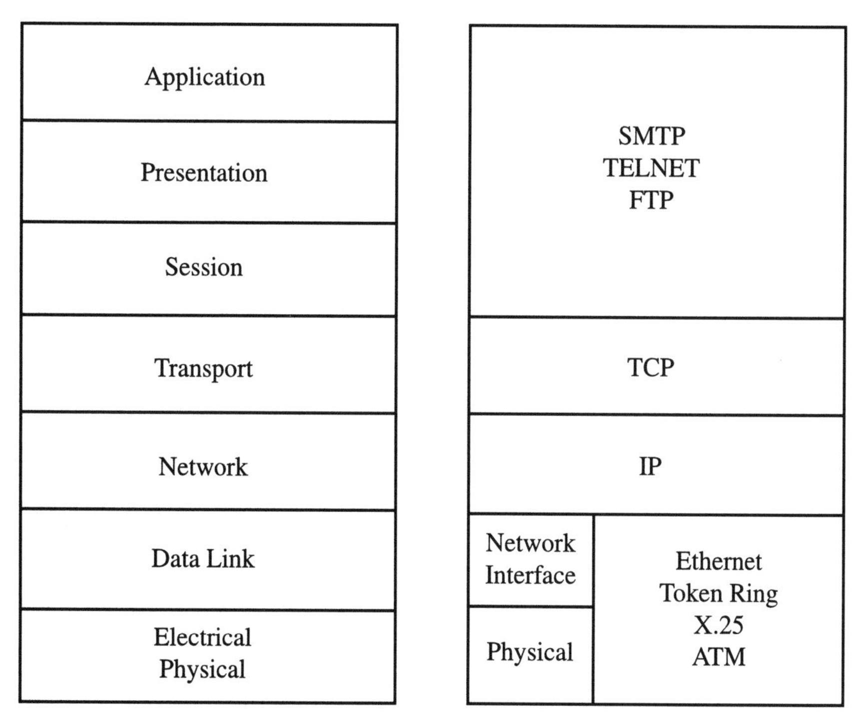

FIGURE 2.4 OSI protocol and TCP/IP protocol.

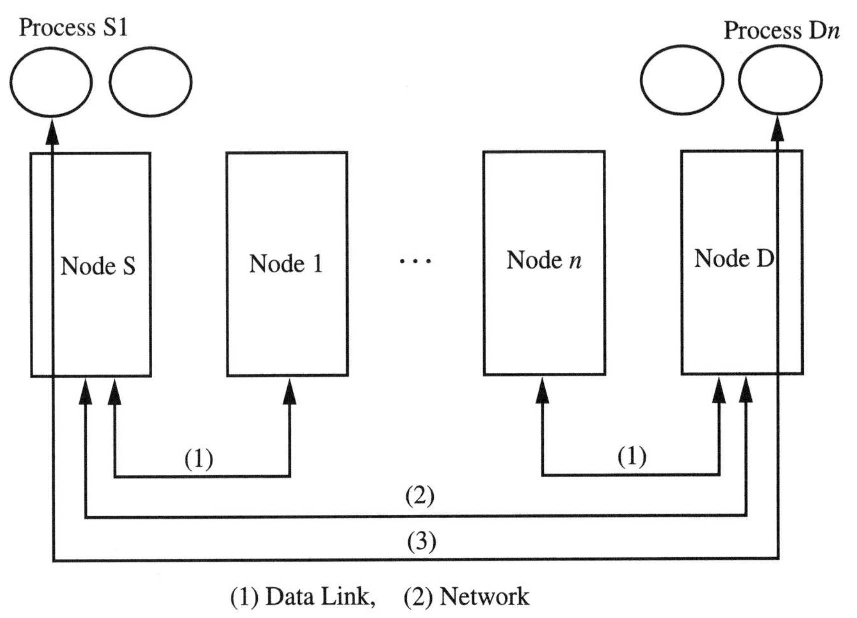

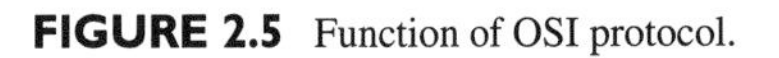

FIGURE 2.5 Function of OSI protocol.

transmission both ways) or half-duplex transmission. This is done via a session-level protocol. A presentation-level protocol provides the code conversion for the message. An application-level protocol provides the file transfer function, job transfer function, or telnet remote access function. These protocols have been standardized, and they are called OSI standard protocol. The TCP/IP protocol is used on the Internet. The IP protocol corresponds to the OSI network–level protocol. The TCP protocol corresponds to the OSI transport–level protocol. The TCP/IP protocol is shown in Figure 2.4. The OSI protocol is presented in Figures 2.4 and 2.5. Further details of the OSI and TCP/IP protocols are described in later chapters.

3

COMMUNICATION NETWORK STRUCTURE

The telephone network was constructed as a communication network. Using the network, telephone service was provided. Advances in information processing technology and new communication services such as data communication service and facsimile communication service have been implemented by using the telephone network. To interconnect terminals, computers, and networks, the network architecture and protocol have been developed. In this chapter, the network architecture of the telephone network is described, as is the network architecture and protocol of the computer network.

3.1 TELEPHONE NETWORK ARCHITECTURE

The conventional telephone network structure is shown in Figure 3.1. The conventional telephone network has been hierarchically structured. There are toll switching (TS) systems and local switching (LS) systems. When phone A calls phone B, the call is transmitted through LS-TS-TS-TS-TS-LS. With advances in telecommunication network, the network structure became simple. A two-layered structure

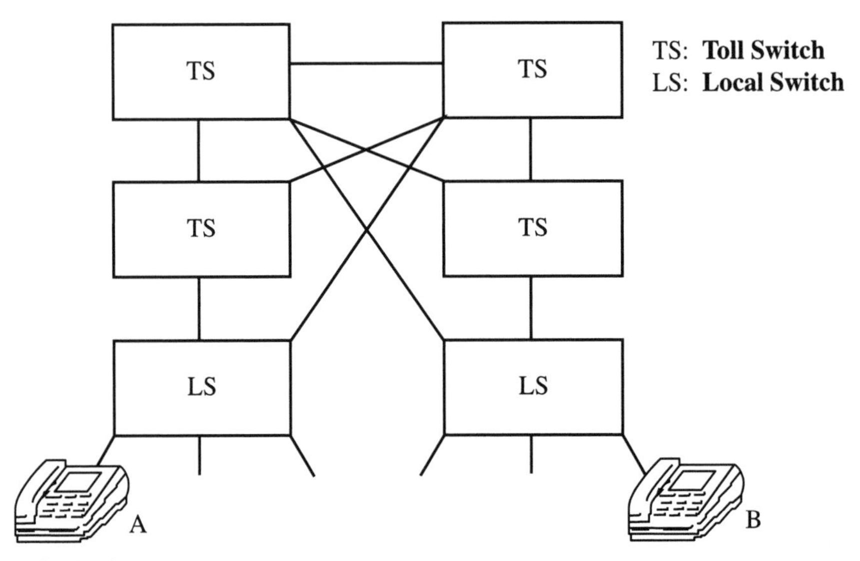

FIGURE 3.1 Communication network structure.

including LS and TS has been used in Japan in its public telephone network. When the line is busy, an alternate line is selected and the call is transmitted through an alternate route. This is called *routing*.

Recently, it has become increasingly necessary to transmit multimedia information in real time via communication lines. To accomplish this, high-speed transmission lines or digital transmission lines have been constructed. The high-speed and digital transmission networks are called the *Information Superhighway*. The high-speed network project has been conducted in the public telephone network.

As shown in Figure 3.1, first the transmission lines between TSs are digitized by the introduction of optical-fiber links. Second, the links between TS and LS are digitized. Finally, subscriber lines are digitized with a capacity of 156 Mbps. Multimedia information can be transmitted via subscriber lines in real time. Through the introduction of high-speed networks, video signals and motion pictures can readily be sent to subscribers in real time.

3.2 COMPUTER NETWORK ARCHITECTURE

In this section, computer network structure, computer network architecture, and the OSI protocol are described.

A computer network is composed of networks and computers. In order to interconnect different types of computers, a network architecture and protocols are standardized. For example, the OSI reference model has been established to interconnect heterogeneous networks. After development of the OSI model, the TCP/IP protocol was developed and put into practical use for local area networks.

3.2.1 Computer Network

A computer network is composed of networks and computers and is used to interconnect computers that are widely distributed. The computer network contributes to the functionality, usability, reliability, and efficiency of the distributed computers.

In the first step, the centralized computer system was developed. Here, a central computer and terminals are linked and various kinds of application programs are provided, such as inventory management and process control. In the second step, the distributed computer network was developed. In this system, two or more distributed computers are interconnected via the network, with terminals linked to each computer. Each computer has its own functions, such as inventory management and database management. A request from a terminal is distributed to the computer, which fulfills its request.

The Advanced Research Project Agency (ARPA) is a typical example of a distributed computer network. The ARPA project, which started in 1969 under the sponsorship of the U.S. Department of Defense, interconnected the computers of universities and research institutes. Out of this activity, the packet switched network, the TCP/IP protocol, and the network architecture were invented.

The ARPA network has become the backbone of the Internet. The TCP/IP protocol, which was invented by the project, has been used as the de facto standard of the Internet. Mail message–handling services, such as electronic mail, and bulletin boards were introduced in the ARPA network.

3.2.2 Network Architecture

A computer and a terminal exchange information via the computer network. Therefore, the network should be efficient, fast, and reliable with respect to the transmission of data. To transmit data between computer and terminal, between peer computers, and between networks, protocols have to be developed and standardized. All kinds of computers and terminals should be linkable in the network. The network is also expanded by interconnecting to other networks. In this way, the network structure will change dynamically day by day.

To achieve this, the network architecture—such as protocols and network topology—has to be defined and standardized. The network architecture should be such that any kinds of components, such as terminals, computers, and networks, can be interconnected without any restrictions.

ARPA was the network architecture invented first in the world. Since then, computer manufacturers such as IBM, Digital Equipment, Hitachi, Fujitsu, and NEC have announced their own architecture for their computer network. For example, SNA was the architecture developed by IBM, and DECNET was developed by Digital Equipment.

These architectures differed from each other, so standardization was proposed and conducted, mainly by ISO and CCITT. The OSI reference model was proposed and standardized as an architecture for computer networks. When making standard protocols, the following points are taken into consideration.

(1) Information must be transmitted properly. It is necessary to specify the interface between a terminal and a line linked to the network. For example, the physical conditions, such as electricity and the connection between a terminal and a line, should be specified. It is also necessary to specify error detection and recovery during data transmission and to specify the sequence control and flow control.

(2) Information must be processed properly. It is necessary to specify how to transfer data between a terminal and a computer, the particular character set to be processed between a terminal and a computer or between peer terminals or peer computers, and the data format and commands to be processed in the network.

3.2.3 OSI Protocol

There are three logical components in OSI: application process, open system, and transmission medium. The *application process* is the process conducted in a terminal or in a computer. *Open system* is a platform that provides the information processing and communication function between peer application processes. The transmission medium is a line that transmits information and signals between open systems. Open system provides the functions for interconnecting two or more systems and includes such equipment as a terminal or a workstation and a network in which terminals and computers are interconnected.

In the OSI reference model, seven layer protocols are defined, from a physical-level protocol to an application-level protocol. The lower-level protocols, such as a physical-level protocol and a data link–level protocol, define the functions of the communications hardware. The upper-level protocols, such as an application-level protocol and a presentation-level protocol, define the functions of communication processing. The protocols are a set of communication functions between peer nodes, that is, the interface between them.

The protocols are well defined to ensure the transparency of the interconnection between peer entities and between neighboring layers. An upper-level protocol issues a request for the communication functions provided by the adjacent lower-level layer. The adjacent lower-level layer provides its functions to the adjacent upper-level layer, although it does not control the adjacent upper-level layer. Figure 3.2 shows the layered structure of the OSI protocol. In it, the $(N + 1)$th layer is the adjacent upper-level layer of the Nth layer.

The OSI reference model is composed of seven layers. The functions of the Nth layer are composed of the entity, service, and protocol of the Nth layer. The Nth entity creates the Nth service by using the $(N - 1)$th entity. The Nth service is provided to the $(N + 1)$th entity.

The Nth service is divided into connection-type service and connectionless-type service. In case of the connection-type service, a connection is established between a source node and its destination node before the data transmission begins. After finishing the data transmission, the communication link is disconnected.

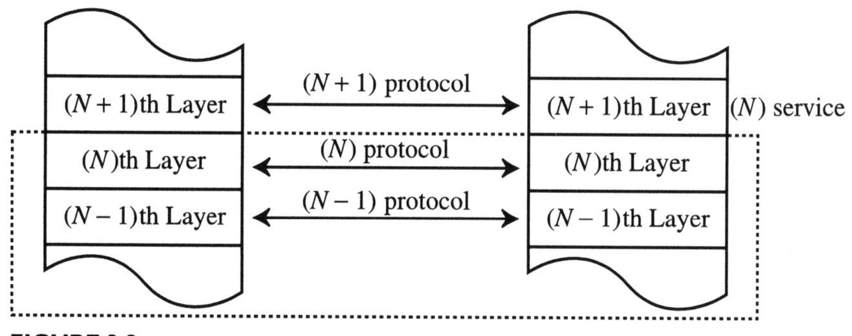

FIGURE 3.2 Layered structure of the OSI protocol.

A virtual circuit of the packet switching system is an example of connection-type service.

On the other hand, in the case of the connectionless-type service, the *N*th entity is a functional module for the communication between a source and its destination. The entity has the functions for communication between peer nodes and the functions for communication between the entity and the adjacent upper-level entity or between the entity and the adjacent lower-level entity.

The *N*th service provides the communication functions to the $(N + 1)$th entity. Generally speaking, the *N*th entity provides the *N*th service to the $(N + 1)$th entity by using the $(N - 1)$th service provided by the $(N - 1)$th entity in cooperation with the peer *N*th entity. The access point in which the $(N + 1)$th entity receives the *N*th service is defined as the *N*th service access point (SAP). The information exchanged through the *N*th SAP is defined as the *N*th service primitive.

The *N*th connection is a communication channel between the *N*th entity and the peer *N*th entity. The channel is used for data transmission between the $(N + 1)$th entity and the peer $(N + 1)$th entity. The *N*th connection is given a specific identifier. The identifier is attached to the transmitting data. Therefore the *N*th entity can send the data to the $(N + 1)$th entity by recognizing the identifier.

The *N*th protocol is defined as the protocol by which the *N*th entity communicates with the peer *N*th entity. In the protocols, there are the protocols for establishment of the connection, information control, and other necessary actions.

The unit of the data block in the *N*th layer is defined as the *N*th protocol data unit (PDU). As shown in Figure 3.3, the $(N + 1)$th PDU is manipulated as the *N*th service data unit (SDU). Basically, the $(N + 1)$th PDU is replaced by the *N*th SDU. According to the data length, more than one $(N + 1)$th PDU are integrated into a single *N*th SDU. The *N*th PDU is created by attaching the *N*th protocol control identifier (PCI) to the *N*th SDU.

Generally speaking, the current layer control information is attached to the adjacent upper-level layer PDU to generate the current layer SDU. Figure 3.3 shows the relationship between PDU and SDU.

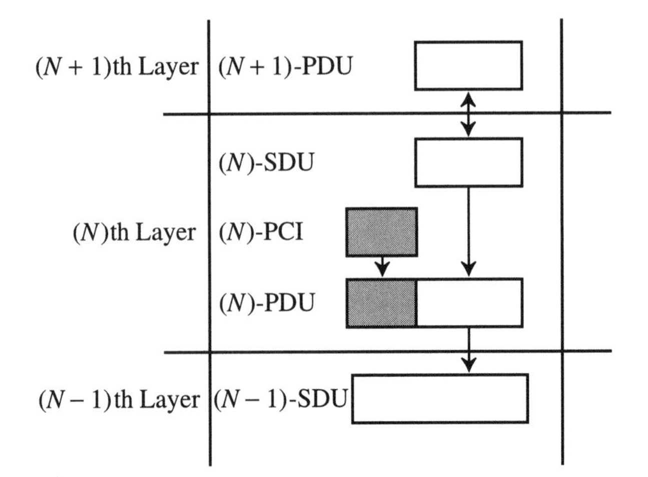

FIGURE 3.3 Structure of the data unit.

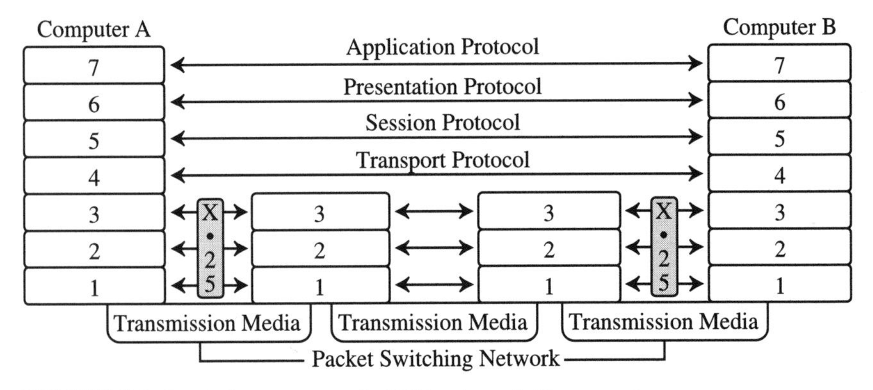

FIGURE 3.4 Protocol structure over the X.25 packet switching network.

3.2.4 Specific Structure of the OSI Reference Model

In the OSI reference model, a seven-layer model is specified. Each layer has its own specific function, is independent of any other layer, and has an interface with adjacent layers. The seven-layer model is shown in Figure 3.4, which portrays a node and the peer node. Each node, which may be a computer, a terminal, or a workstation, has seven layers: a physical layer, a data link layer, a network layer, a transport layer, a session layer, a presentation layer, and an application layer.

The functions of each layer are as follows. The first layer, the physical layer, defines the rules and interfaces of the bit streams transmitted between adjacent nodes. The electrical, mechanical, or physical conditions, such as the electrical current, voltage, or pin size or its layout, are defined in this layer. Specifications such as RS-232C, RS-422/423, RS-449, and X.21 are examples of this layer.

The second layer, the data link layer, defines the procedures of connection or disconnection and the transmission control between adjacent nodes. It contributes to the precise, efficient, and prompt transmission between adjacent nodes. It has the functions of error correction control, a sequence control, and a flow control. Examples of this layer are the basic transmission control, high-level data link control (HDLC), and LAPB.LAPD of ISDN or logical link control (LLC) of a local area network. Media access control (MAC) protocols of the local area network, such as CSMA/CD, token bus, and token ring, are included in the data link layer and the physical layer.

The third layer is a network layer. Using the functions of this layer, a transparent transmission path is established between a source and its destination. The functions include flow control, routing, and sequence control, providing for precise and speedy data transmission control throughout the network. Examples of these functions include the X.25 protocol, which is the user network interface of the packet switching network, and the X.75 protocol, which is a network-to-network interface.

The fourth layer is a transport layer. Using the functions of this layer, a source process and its destination process are linked and a transmission path established between the source process and its destination process to communicate together. This layer is also called the end-to-end transmission layer. It provides the functions of flow control, sequence control, the composition and decomposition of data, and the detection of data loss during transmission.

There are five classes in this fourth layer. Class 0 provides functions such as connection establishment between peer processes, data composition or decomposition, or transmission of the transport protocol data unit (TPDU). The higher the class, the more advanced the functions that are accommodated. For example, class 4 provides not only the basic functions but also the advanced functions, such as flow control, sequence control, multiplication, and error check and control, in order to support high-quality transmission even over low-quality transmission lines.

The fifth layer is a session layer. It provides the conversational functions between the adjacent entities of a presentation layer. Namely, this layer provides the functions by which the connection called *session* is established, maintained, and released. Additionally it provides some kinds of conversation styles and checking functions for error recovery.

The sixth layer is a presentation layer. It provides the data conversion facilities to the application programs or terminals in an application layer. The services include code or character conversion, data form or layout conversion, data compression of images, and encryption/decryption for security.

The seventh layer is an application layer. It provides a client with application programs for accessing the OSI environment. In the application layer are functions for file transfer, job transfer, virtual terminal, database access, transaction processing, or the mail handling system (MHS).

The OSI reference model has been used for other standardization activities of computer communication or data communication since it was established. An example of a protocol structure on the packet switching network is shown in Figure 3.4.

The OSI reference model was implemented at the U.S. National Bureau of Standards as OSINET. The Manufacturing Automation Protocol (MAP) was implemented by the consortium of GM and Boeing.

3.3 INTERNET NETWORK ARCHITECTURE

On the Internet, local area networks (LANs) are interconnected via dedicated lines or telephone networks. A local area network is installed in the intraoffice network. In LAN, terminals such as workstations, personal computers, and/or computers as file server or database server or mail server are linked to the bus or ring network. The network provides 1.5–100Mbps transmission lines. The Internet is called the network of networks. Local area networks are interconnected by telephone network or dedicated lines to form the Internet.

There are ring and bus network structures in the local area network topology. Using the Internet, various kinds of IT services are provided, including mail handling, database access, file access, continuous acquisition and lifelong support (CALS), electronic payment, and electronic commerce services. Security is introduced to protect information transmitted over the Internet from hackers, dishonest users, and wiretapping. Encryption and decryption are implemented to build secure networks.

To achieve the high-speed Internet, advanced Internet projects, such as Internet 2, are going on.

3.3.1 TCP/IP Protocol

TCP/IP has been widely used in the Internet as a de facto standard protocol. The TCP/IP protocol initially was developed as ARPA network protocols and was improved to include the concept of network architecture. It was implemented as the standard protocol in Unix 4.2 Berkeley software distribution (BSD). Because it was used in the Unix operating system, it has been widely employed on the Internet, where Unix was mainly used. The TCP/IP protocol is shown in Figure 2.4, where it is compared with the OSI protocol.

In accordance with the advances on the Internet, workstations that have a reduced instruction set computer (RISC) and run on the Unix operating system have been developed and put into practical use. At the same time, local area networks have been widely used and have expanded throughout the globe. In order to interconnect workstations, the TCP/IP protocol has been used as their standard protocol. Because TCP/IP has been used mainly in local area networks, it has been recognized as the de facto standard protocol in the local area network environment. The TCP/IP layer structure is shown in Figure 2.4. There are five layers from bottom to top: physical, network interface, Internet, transport, and application.

The application layer includes such application programs as the Telecommunication Network Protocol (TELNET), the File Transfer Protocol (FTP), and the Simple Mail Transfer Protocol (SMTP).

The transport layer provides end-to-end communication between adjacent application programs. It decomposes the data transferred from the application layer and creates the TP (Transport Protocol) packet, attached with the control information, such as a program identifier. Then it transfers the TP packet to the Internet layer.

The Internet layer provides the communication functions between a source computer and its destination one. It receives a TP packet and the destination IP address. Then it constructs the IP datagram using the TP packet and the destination IP address. Using the routing algorithm, it decides the destination computer or the gateway processor and transfers the IP datagram to the network interface layer.

The network interface layer provides the control and interface functions for transmitting the IP datagram through the physical layer. To achieve this, it creates an HDLC frame or LAN frame, depending on the physical network structure.

When the physical layer is a LAN structure, it corresponds to a device driver or LAN interfacer. When it is a public network, such as the packet switched network, it corresponds to the communication equipment based on the X.25 standard.

TCP/IP has been implemented by clients independent of OSI activities. Therefore, it does not match the OSI structure. Basically, the application layer corresponds to OSI layers 5 to 7. The transport layer corresponds to OSI layer 4. The internet layer is like OSI layer 3. The network layer corresponds to OSI layer 2. The physical layer is like OSI layer 1.

3.3.2 TCP/IP Subprotocol Structure

Each layer of the TCP/IP is composed of a set of subprotocols that correspond to entities of the OSI reference model, as shown in Table 3.1. The application layer provides the protocols that users directly access. In the application layer, which is the top layer, are subprotocols such as the Simple Mail Transfer Protocol (SMTP),

TABLE 3.1 TCP/IP Subprotocols

<table>
<tr><th>Layer</th><th colspan="7">Subprotocols</th></tr>
<tr><td>Application Layer</td><td>SMTP</td><td>FTP</td><td>TELNET</td><td>DNS</td><td colspan="2">TFTP</td><td rowspan="2">NVP</td></tr>
<tr><td>Transport Layer</td><td colspan="4">TCP</td><td colspan="2">UDP</td></tr>
<tr><td rowspan="2">Internet Layer</td><td colspan="7">ICMP</td></tr>
<tr><td colspan="3">IP</td><td colspan="2">ARP</td><td colspan="2">RARP</td></tr>
<tr><td>Network Interface Layer</td><td colspan="2" rowspan="2">Ethernet (CSMA/CD LAN)</td><td rowspan="2">Internet</td><td colspan="2" rowspan="2">X.25 Packet Switching Network</td><td colspan="2" rowspan="2">Others</td></tr>
<tr><td>Physical Layer</td></tr>
</table>

the Domain Name Service (DNS), Telecommunication Network Protocol (TELNET), and File Transfer Protocol (FTP). SMTP is the protocol that provides message transfer functions between computers. It is used for electronic mail and bulletin board services. DNS provides the service that translates a domain name to the IP address. TELNET is the protocol that establishes the TCP connection between a user's computer and a remote peer computer. Through this, he or she can issue a remote login and access the remote computer. FTP is the protocol that provides the file transfer between computers. Using FTP, a user can log onto a remote computer, access the directory of the file, and copy the contents of the file. The connection is established by TELNET before FTP is used. In this layer, are the Trivial FTP (TFTP) as the simple file transfer protocol and the Network Voice Protocol (NVP) as the protocol for voice transmission.

In the transfer layer is the Transport Control Protocol (TCP), which enables connection-type communication between two nodes. It corresponds to the virtual circuit on a packet switching system and is a typical protocol on the Internet. The User Datagram Protocol (UDP) provides connectionless-type communication and corresponds to the datagram communication in the packet switching network.

Table 3.1 shows the dependence among the subprotocols, both in the application layer and in the transport layer. For example FTP and TELNET use TCP, and TFTP uses UDP.

In the Internet layer, IP is the fundamental protocol. It provides a connectionless-type data transmission function between a node and its peer node via a number of communication networks. IP specifies the format of the IP datagram, how to perform a routing, and how to correct errors. The Internet Control Message Protocol (ICMP) is the protocol that transmits the control information concerning the monitoring of communication networks or gateways between a node computer and its peer computer. The Address Resolution Protocol (ARP) or the Remote ARP (RARP) is the protocol that translates an IP address to its physical address on the Ethernet, and vice versa if needed.

Both the network interface layer and the physical layer specify the communication networks for data transmission, such as the Ethernet, ARPA network, and X.25 packet switching network, and the interface.

4

ADVANCES IN COMMUNICATION NETWORKS

The telephone network is used mainly for transmission of voice signals, which are analog. Frequency modulation or amplitude modulation is used for transmission of voice signals. The telephone network is an analog network with a transmission capacity of 3.4 kHz, which is necessary to transmit voice signals. About 30 years ago, the data communication system was developed using the telephone network or dedicated lines. In this system computers are interconnected through the telephone network or dedicated lines to transmit information. This was followed by the facsimile communication system and the videotex communication system, which were developed to transmit facsimile and video signals, respectively. These systems developed separately; it is therefore costly to construct or maintain them.

With the data communication system, facsimile communication system, and videotex communication system, information to be transmitted is digitized. In the telephone network, the voice is an analog signal. When the voice signal is digitized, all of the foregoing kinds of information can be transmitted on digital lines. In this way, any kind of communication service can be provided via a single digital network. This idea is called the integrated services digital network (ISDN).

Prior to ISDN, each service was provided separately through its own network. With ISDN all services can be transmitted via a single digital network. Using ISDN, the following new service can be achieved. When the phone rings we don't know who is calling. We know who is calling *after* we hang up. This is a source-oriented communication service. Thanks to the introduction of the digital network, we can now have 2B + D channels in a subscriber line. By 2B we mean two base-band channels; D means one data channel. By using the D channel we can transmit information to identify the ID of the source phone number. So when the phone rings, the source ID can be shown on the display of the telephone before we answer the call. This gives us a choice of whether or not to answer the call. We call this a destination-oriented communication service. Through the introduction of ISDN, a more human-friendly telephone service is achieved.

4.1 INTEGRATED SERVICES DIGITAL NETWORK

In this chapter, ISDN is described. To provide the wide area network service, it is necessary to provide network architectures for wide area. There are N-ISDN, B-ISDN, and ATM switching systems for this purpose.

As multimedia services evolved, it became necessary to transmit not only voice signals but also video, image, and text information through the network concurrently. For this purpose, ISDN architecture was proposed and implemented. ISDN provides the transmission of all kinds of data through a single channel in time-division mode. Depending on the transmission speed, ISDN is classified as N-ISDN or B-ISDN.

4.2 N-ISDN

N-ISDN has been standardized as I series recommendations by ITU-T. I series consist of I.100, I.200, I.300, I.400, I.500, and I.600. I.100 defines the basic concepts of ISDN. I.200 defines the service specifications of ISDN. I.300 specifies the network functions of ISDN. I.400 specifies the interface of user and network. I.500 specifies the internetworking interface. I.600 specifies the maintenance and management functions of ISDN.

As user–network interface (UNI) reference points, points T, S, and R are specified, as shown in Figure 4.1. Point T is the terminal point of the network as well as the interface point of network terminal equipment NT1. When NT2, such as a PBX or an LAN, is connected to point T, the terminal point of NT2 is called point S. TE1, such as a digital telephone, a G4 fax, or digital equipment, is connected to point S. In the case of analog equipment TE2, such as an analog telephone, or an analog fax, the interface equipment TA acts as the interface between TE2 and NT2. The interface point between TA and TE2 is point R. In N-ISDN, the B-channel, D-channel, and H-channel are provided. The B-channel is a user's communication

TABLE 4.1 Type of Channel in N-ISDN

Type of Channel			Speed of Channel
B			64 kbps
D			16 kbps 64 kbps
H	H_0		384 kbps
	H_1	H_{11}	1536 kbps
		H_{12}	1920 kbps

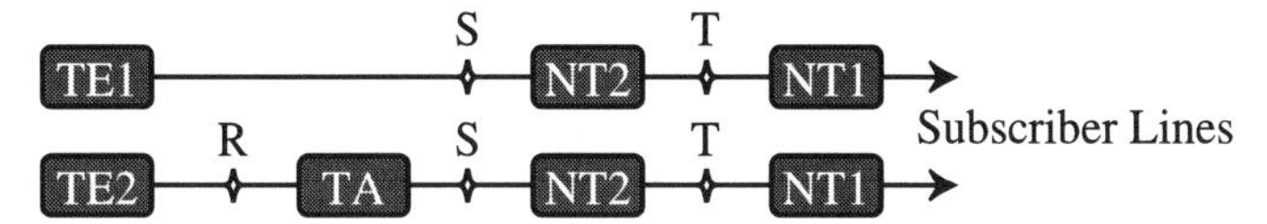

FIGURE 4.1 Terminal-line interface in N-ISDN.

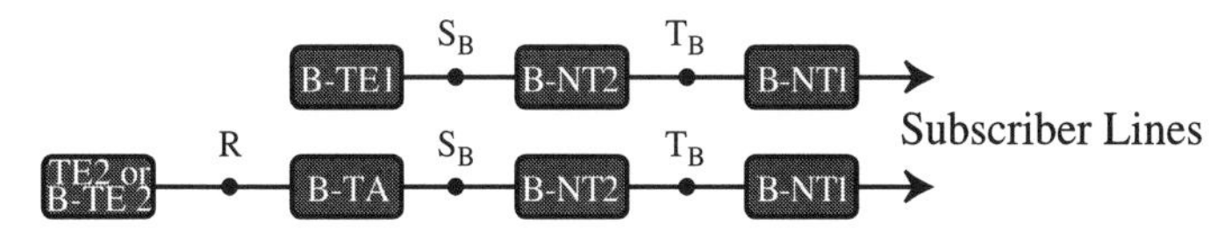

FIGURE 4.2 Terminal-line interface in B-ISDN.

channel, which provides 8-, 16-, 32-, and 64-Kbps transmission services. The D-channel is a control channel and provides 16- and 64-Kbps services. The H-channel is a user's communication channel and provides 384-, 1536-, and 1920-Kbps services. The B-, D-, and H-channel characteristics are shown in Table 4.1.

4.3 B-ISDN

B-ISDN provides the high-speed transmission of digital data through the network. With B-ISDN, there are services such as an interactive service and a distribution service. In the interactive service, a conversational service, a message handling service, and an information retrieval service are provided. In the distribution service, a broadcasting service, such as a radio or a television service, is provided. The UNI of B-ISDN is specified as in Figure 4.2. Points T_B, S_B, and R are specified in this figure. For example, analog equipment TE2, such as an analog television, is connected to B-NT2 via B-TA. Digital equipment TE_1, such as a digital television, is connected directly to B-NT2.

4.4 ASYNCHRONOUS TRANSFER MODE

Asynchronous transfer mode (ATM) provides high-speed switching functions. An ATM packet is called an ATM cell of 53 octets, which consists of control information and data. In a packet switching system, the packet size is variable. Therefore it takes time to identify the packet size and process it. In an ATM switching system, a packet size is 53 octets. Therefore it is easy to identify and process.

The network where switching systems, terminals, and transmission lines are linked is called a *network topology*. Graph theory is used to solve problems concerning network topology. A *node* corresponds to a switching system or a terminal. A *branch* corresponds to a transmission line of a network. The transmission line has characteristics such as transmission cost, distance, delay, capacity, and/or malfunction. The characteristics are evaluated and represented as the cost of a branch. The problem of finding a path that has a minimum cost is called the "searching the shortest path" problem. This problem can be solved by using a graph theory.

According to the graph theory, a graph consists of one or more nodes and one or more branches. Sequence $\{p_s, b_{s2}, p_2, b_{23}, \ldots, p_n, b_{ns}, p_s\}$ is called a path, where $b_{s2}, b_{23}, b_{34}, \ldots, b_{ns}$ are directed branches, p_s is a starting node, and p_t is a terminal node. The number of branches is the length of the path.

The path where the same branch passes less than twice is called a *simple* path. The path where the same node passes less than twice is called an *elementary* path. When two paths exist and both of the starting nodes are the same and both of the terminal nodes are the same, the paths organize a *closed* path. When a path where a starting node is the same as a terminal node, the path is a *cycle*. An example of a graph in general is shown in Figure 4.3. An example of a simple path is shown in Figure 4.4. An example of an elementary path is shown in Figure 4.5. An example of a closed path is shown in Figure 4.6. An example of a cycle is shown in Figure 4.7. The algorithm for finding the path(s) from a starting node to a goal node where a graph is given is called the *searching path algorithm*. Now let's take a look at how it works using an example.

Find the path(s) from S to G in Figure 4.8

(1) Nodes A and B, linked by directed branches from S, are chosen and are described in Figure 4.9.

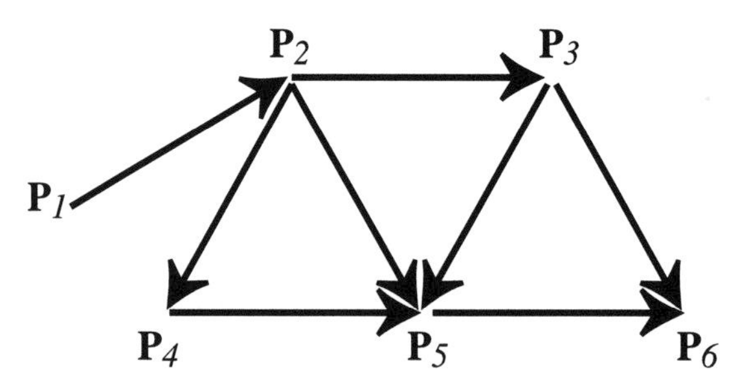

FIGURE 4.3 Graph.

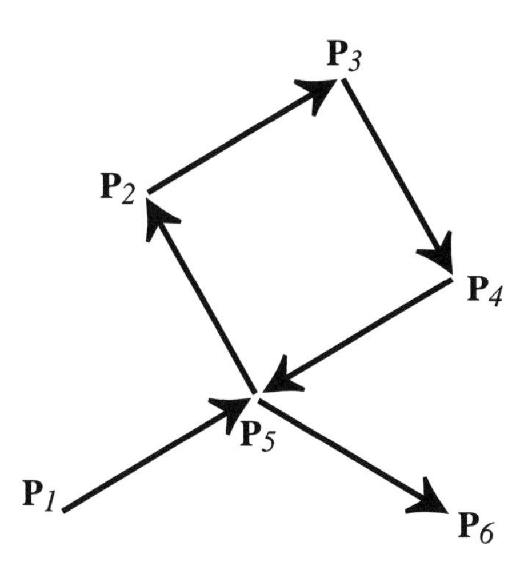

FIGURE 4.4 Simple graph.

FIGURE 4.5 Elementary path.

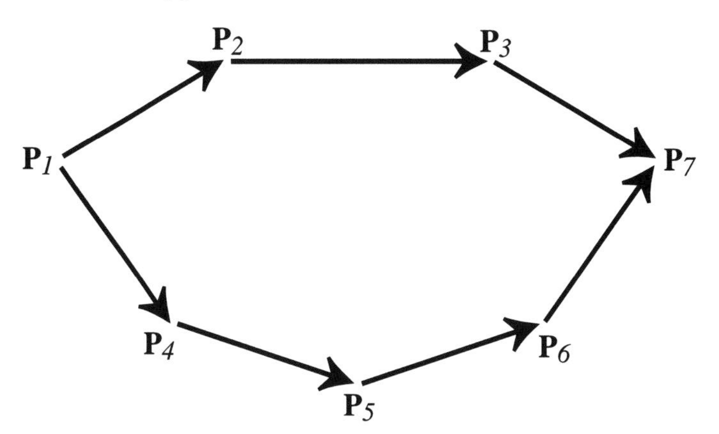

FIGURE 4.6 Closed path.

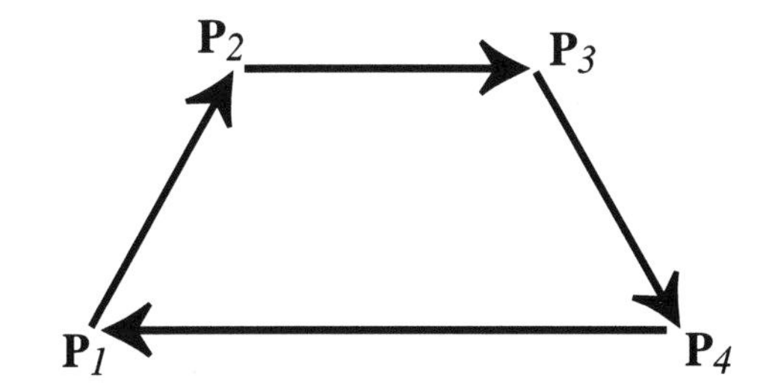

FIGURE 4.7 Cycle.

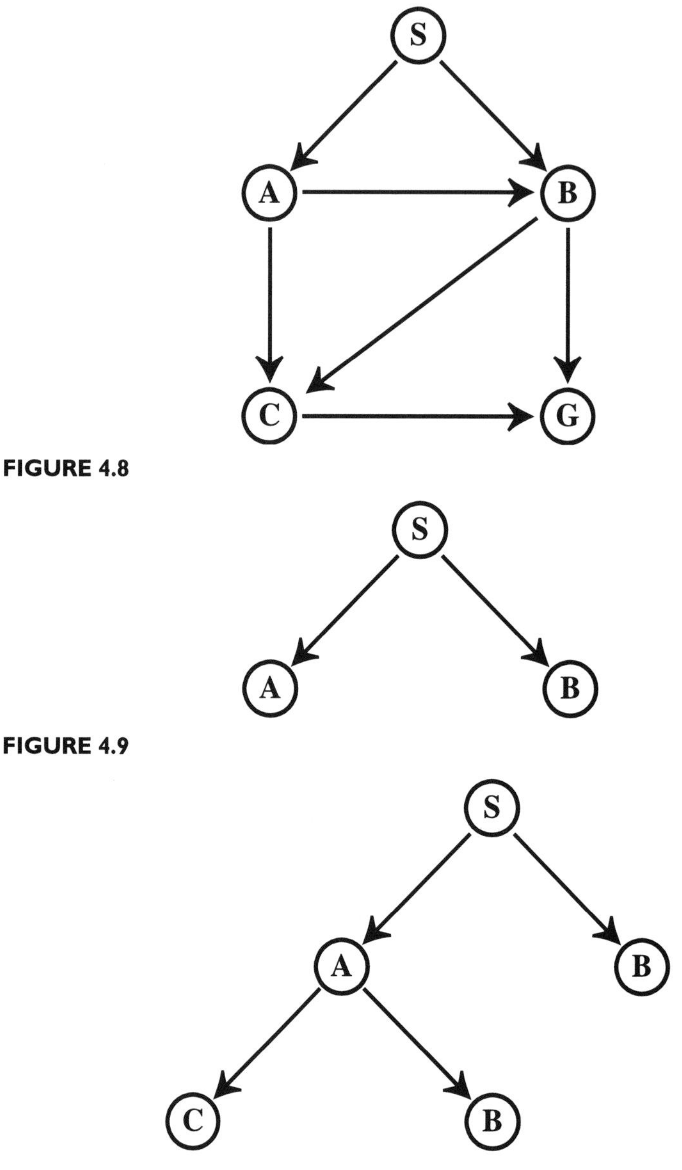

FIGURE 4.8

FIGURE 4.9

FIGURE 4.10

(2) Nodes B and C, linked by directed branches from A, are chosen and are described in Figure 4.10.
(3) Nodes C and G, linked by directed branches from B, are chosen and are described in Figure 4.11.
(4) Node G, linked by directed branches from C, is chosen and is described in Figure 4.12.

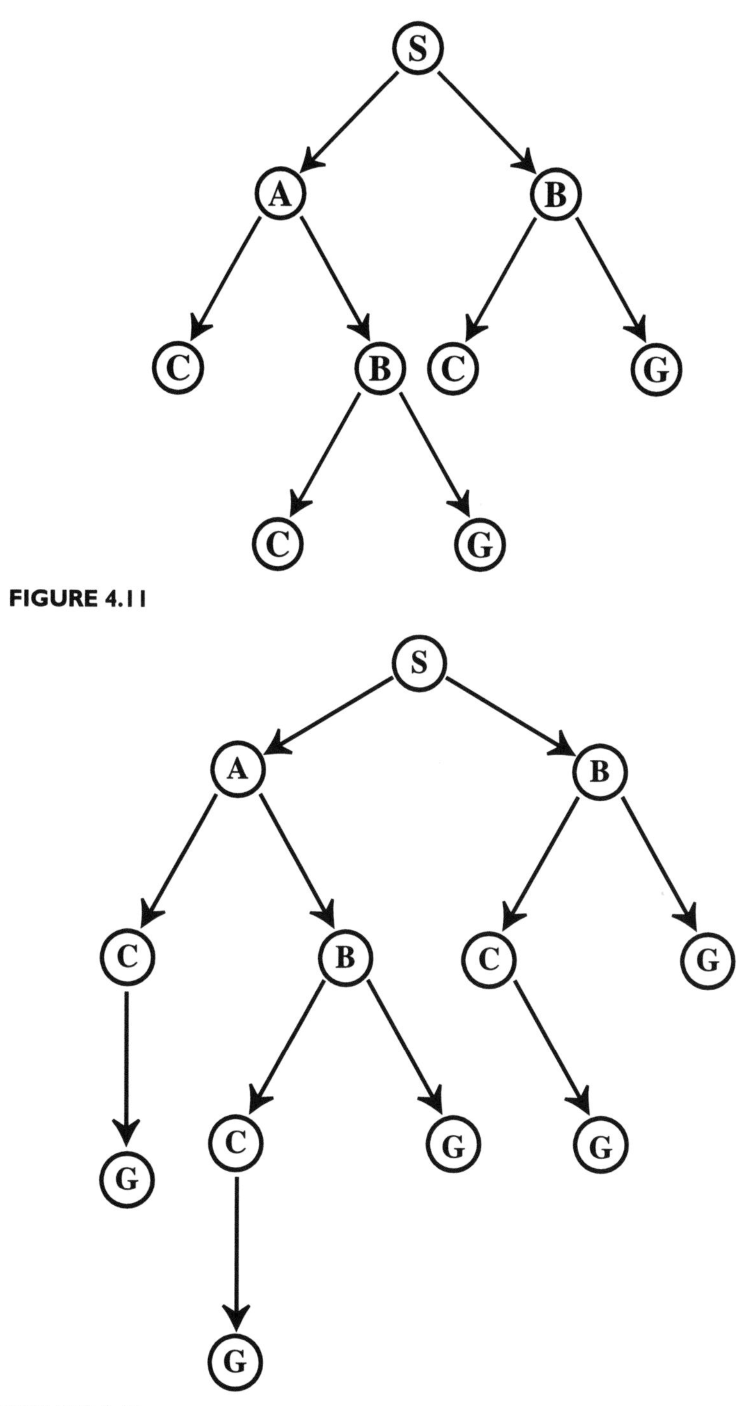

FIGURE 4.11

FIGURE 4.12

According to this analysis, paths from node S to node G are obtained as follows.

(1) S→A→C→G
(2) S→A→B→C→G
(3) S→A→B→G
(4) S→B→C→G
(5) S→B→G

Another example is shown in Figure 4.13. Find a path(s) from node S to node G.

(1) Nodes A and B, directed from S, are chosen (Figure 4.14).
(2) Nodes B and C, directed from A, are chosen (Figure 4.15).
(3) Node D, directed from B, is chosen (Figure 4.16).
(4) Nodes A, C, and G, directed from D, are chosen. In a path {S, A, B, D, A}, the same node A appears twice, the path is a cycle, and then the path is eliminated (Figure 4.17).
(5) Node G, directed from C, is chosen (Figure 4.18).

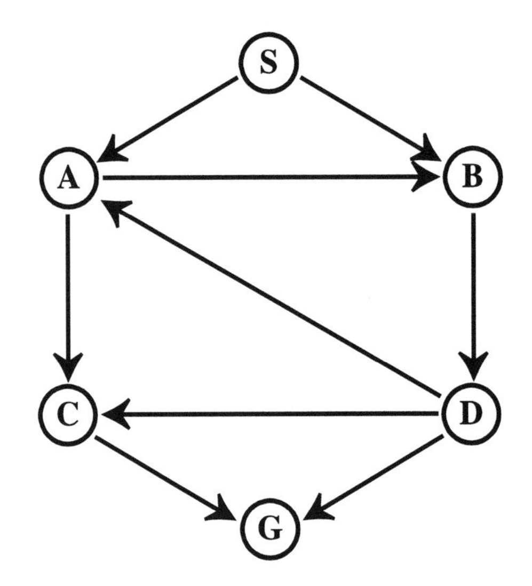

FIGURE 4.13

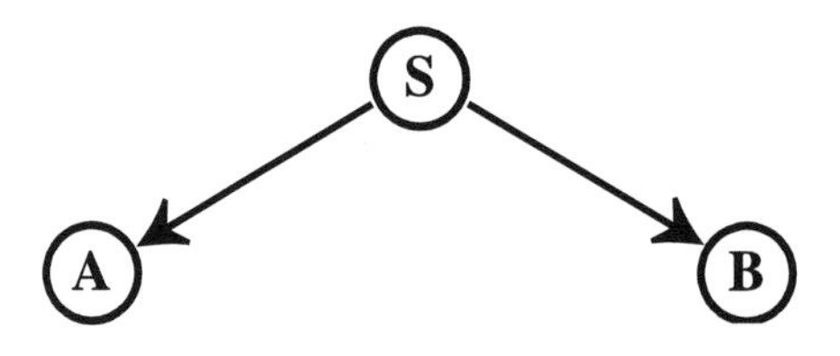

FIGURE 4.14

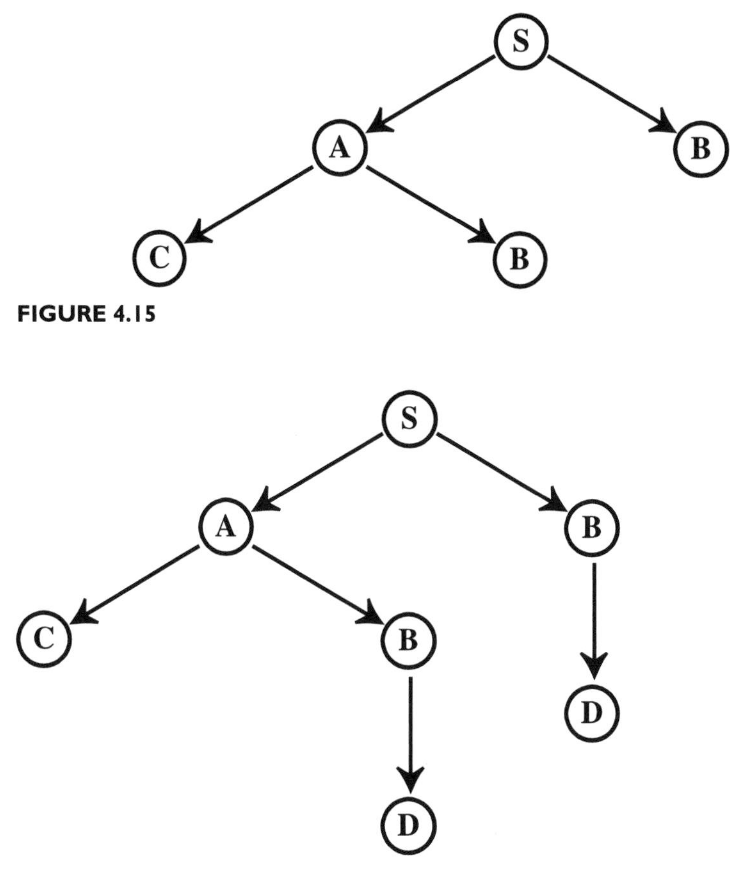

FIGURE 4.15

FIGURE 4.16

(6) Nodes B and C, directed from A, are chosen. In a path {S, B, D, A, B}, node A appears twice and then the path is eliminated (Figure 4.19).
(7) Node G is chosen from C, then the final tree is obtained in Figure 4.20.

The paths directed from node S to node G are as follows.

(1) S→A→C→G
(2) S→A→B→D→C→G
(3) S→A→B→D→G
(4) S→B→D→A→C→G
(5) S→B→D→C→G
(6) S→B→D→G

Now take a look at the searching the shortest-path problem. A graph is given in Figure 4.21. The graph is expanded without cycle into a tree in Figure 4.22.

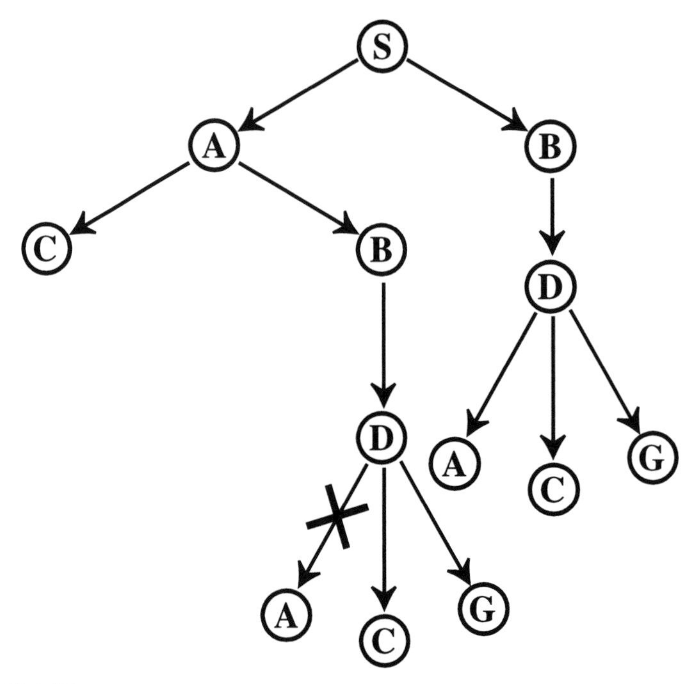

FIGURE 4.17

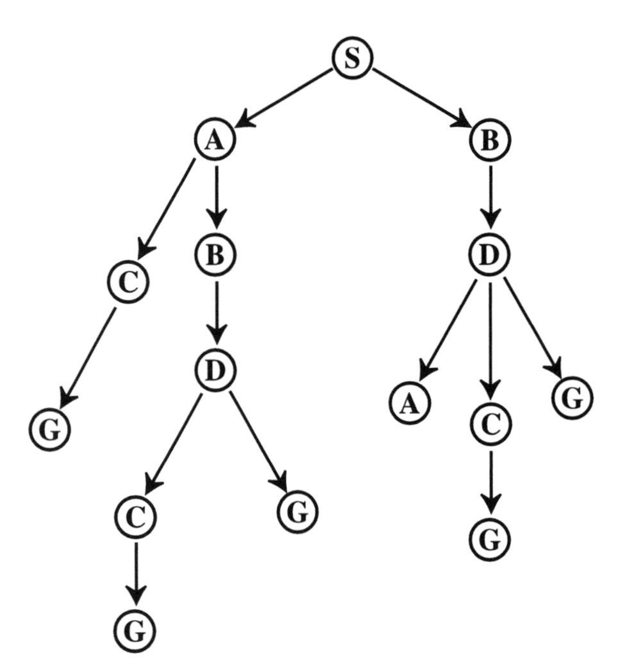

FIGURE 4.18

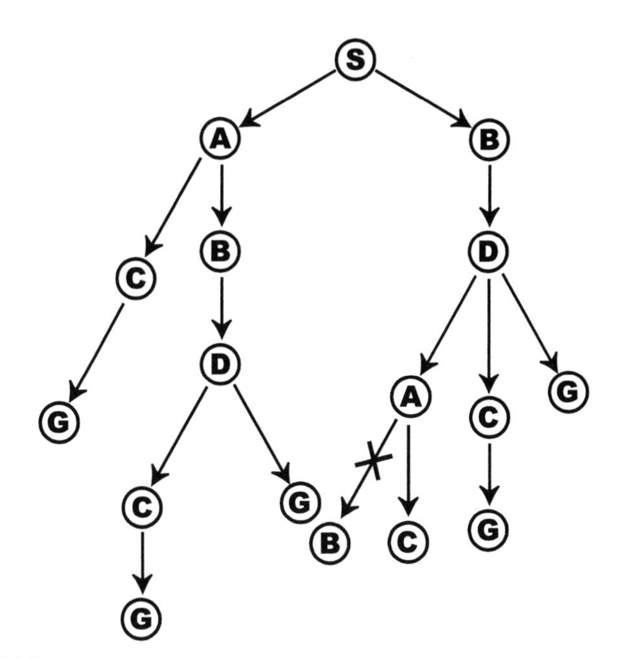

FIGURE 4.19

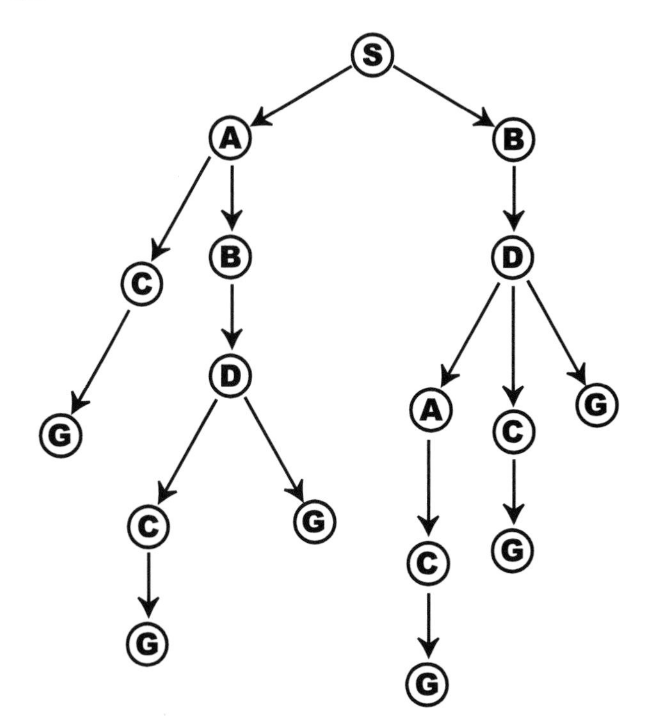

FIGURE 4.20

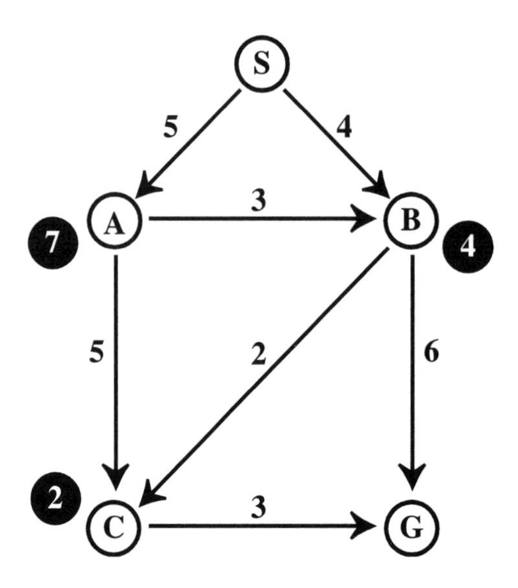

FIGURE 4.21

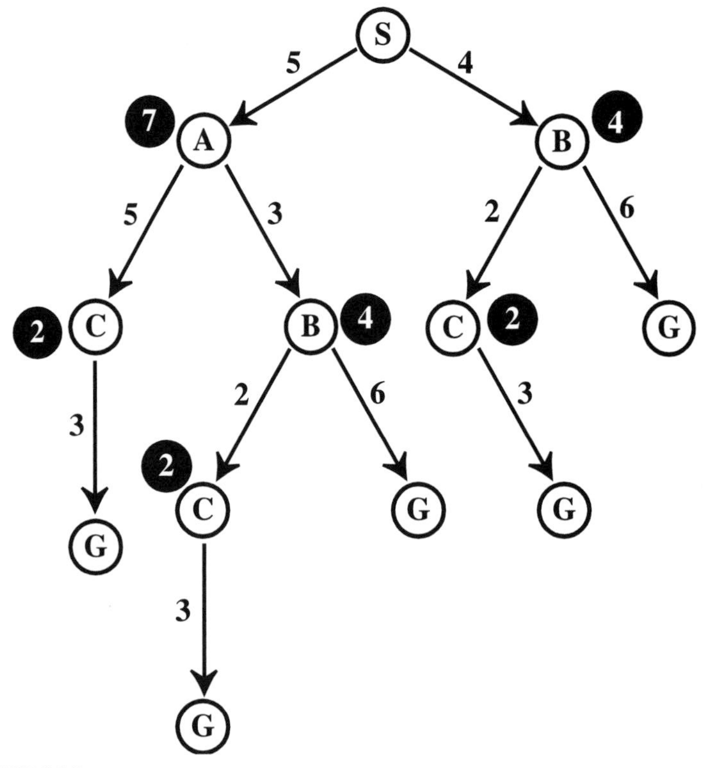

FIGURE 4.22

Consider the shortest path from S to G. A heuristic search method is applied to solve the problem. First of all, an anticipated value is assigned to each node except S. The value AV(N_i→G) is an anticipated value from a node N_i to goal G, where AV is equal to or smaller than the exact minimum value V(N_i-G). In Figure 4.20, an anticipated value is assigned to each node as follows: AV(A→G) is 7. AV(B→G) is 4. AV(C→G) is 2. The processing is performed as follows: The nodes linked by directed branches from S are A and B. The anticipated value of {S→A→>G} is 12, where a value of S to A is 5 and an anticipated value of A to G is 7. Similarly, the anticipated value of {S→B→G} is 8, where the value of S to B is 4 and the value of B to G is 4. Then the shortest path {S→B→G} is chosen. Then the nodes linked by directed branches from B are C and G. The value of {S→B→C→G} is 8, where the value of S to B is 4, the value of B to C is 2, and the anticipated value of C to G is 2. Similarly, the value of {S→B→G} is 10, where the value of S to B is 4 and value of B to G is 6. Then the path {S→B→C→G} is chosen. Finally the value of {S→B→C→G} is calculated and the value is obtained. The shortest path from S to G is {S→B→C→G} and the value is 9.

5

A VARIETY OF TELECOMMUNICATION SYSTEMS

5.1 COMPUTER SHARING

Thirty years ago, computers were very expensive, making it difficult to buy one's own computer. As a result, computer sharing arose, as a system for interconnecting computers and terminals. Users were able to access a computer via a terminal with time-sharing mode. This was a data communication system. The process flow is shown in Figure 5.1

(1) Data is exchanged through path 1, between computer and terminal.
(2) A request for data processing is issued through path 2, from the terminal to the computer.
(3) Data is exchanged between computers through telephone line 3 or dedicated line 4. Data in a local computer is sent to the computer center via networks. High-speed transmission of data is performed via a dedicated line.

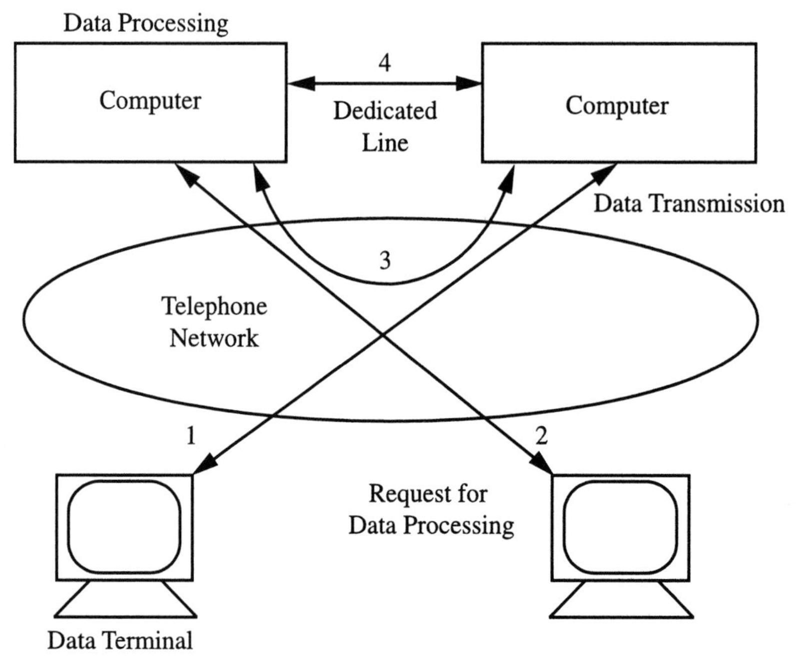

FIGURE 5.1 Data communication system.

5.2 FACSIMILE COMMUNICATION SYSTEM

Facsimile information is transmitted via a telephone line. The facsimile communication system was developed to transmit images at high speed. The structure of the facsimile communication system is shown in Figure 5.2.

(1) TS-FX is a switching system for facsimile information. In Japan the numbering plan for facsimile communication system is 161 or 162.

(2) STOC is a storage system for facsimile information. Facsimile information is stored at STOC temporarily before the information is transmitted via a high-speed dedicated line to another STOC. STOC enables high-speed transmission of facsimile information.

(3) FDIC performs code conversion, media conversion, and speed conversion, for example, character-to-speech conversion or vice versa. Speed conversion is performed when the processing speeds of source terminal and destination terminal are different.

5.3 VIDEOTEX COMMUNICATION SYSTEM

The videotex communication system provides the facility for retrieving image information. The system structure is shown in Figure 5.3.

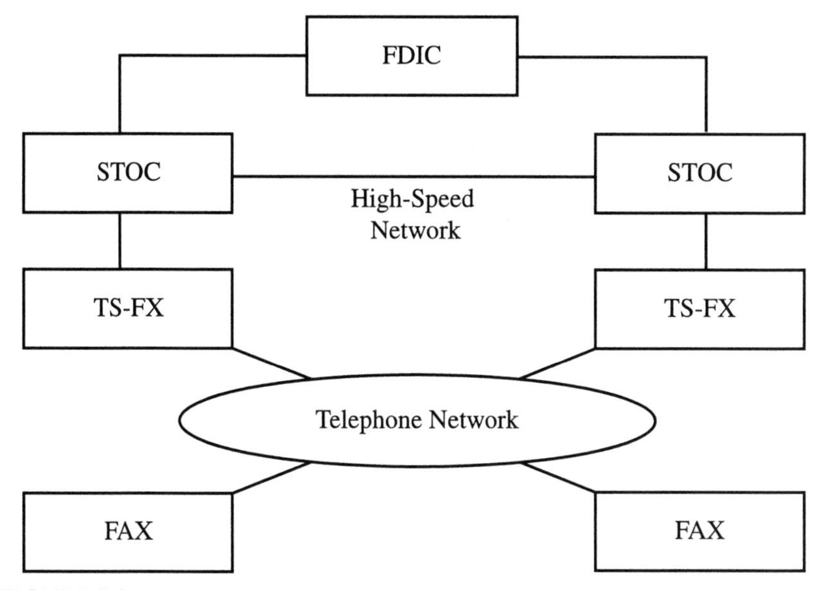

FIGURE 5.2 Facsimile communication system.

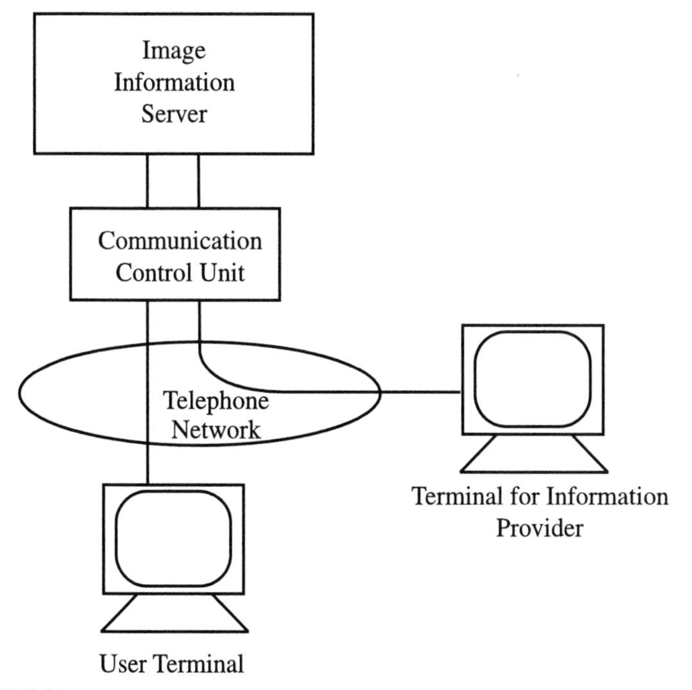

FIGURE 5.3 Videotex communication system.

(1) The image information server retrieves information requested by the terminal from information storage and sends it to the terminal. An image information server stores the information provided by the information provider (IP).
(2) The IP creates image information for the information service and sends it to the image information server via the telephone network.
(3) Users can issue a request for information retrieval from a terminal and receive the information.
(4) The communication control unit controls information or command exchanges among the server, the IP terminal, and the user terminal.

5.4 DISTANCE EDUCATION SYSTEM

Advances in telecommunication technology led to the development of practical distance education systems. To provide education to rural areas with no schools, a distance education system is installed by which lectures are broadcast from a main campus to satellite schools. Students come to the satellite school to attend classes. The main campus and satellite schools are connected via the network, and lectures are transmitted from the main campus to the satellite.

Exchange of lectures among countries and universities is increasing year by year. For example, a Japanese instructor teaches Japanese history to New Zealand students via the system, and a New Zealand instructor teaches the economics of his or her country to Japanese students.

Using the telecommunication networks, many kinds of distance education systems have been developed and put into practical use. Some of them are described here.

5.4.1 Audio Conferencing System

The audio conferencing system uses telephone lines to transmit audio signals. It is less expensive than other distance education systems. Two or more sites are interconnected via the telephone network. Users can hear the transmitted voices but cannot see one another. Instead, pictures of the students attending classes at the remote sites are posted on a board. A diagram of the setup is shown in Figure 5.4.

5.4.2 Audiographic Conferencing System

In the audiographic conferencing system, two telephone lines are used to transmit both audio and graphics. Two or more sites are interconnected via the telephone network. Because it takes a lot of time to transmit text or graphics through a telephone line, it is desirable to download such information *before* the class starts.

FIGURE 5.4 Audio conferencing system.

FIGURE 5.5 Audiographic conferencing system.

Then during the class, students can retrieve information from their terminals in real time. In this system, the students who attend the class at remote places cannot be seen, so their pictures are posted on a board. A diagram of the setup is shown on Figure 5.5.

5.4.3 Video Conferencing System

In the video conferencing system, two channels are used to transmit both audio and video signals. It takes a lot of bandwidth to transmit video signals. Though expensive, this system has been widely used in distance education. Students not only see each other but also hear one another's voices and can see the text or graphics transmitted from the remote site. To reduce the cost, video compression technology has been developed and installed in the system. This includes MPEG1 and MPEG2 video compression technologies. MPEG1 provides 1.5 Mbits

FIGURE 5.6 Video conferencing system.

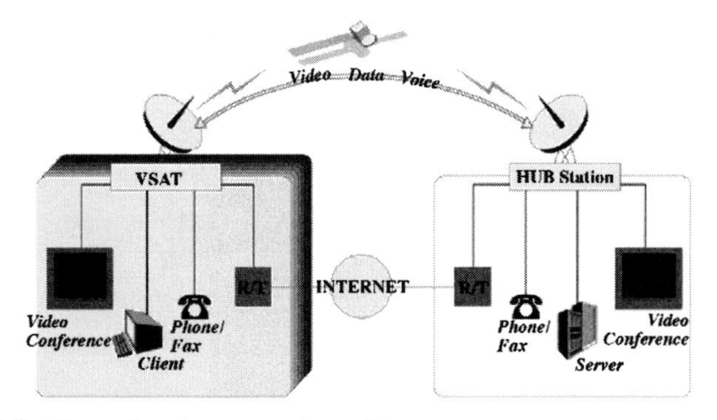

FIGURE 5.7 Distance learning system via satellite.

video transmission and MPEG2 5–6 Mbits. A diagram of the setup is shown in Figure 5.6.

5.4.4 Satellite Communication System

A satellite communication system involves a transmitter and a receiver installed at each station. A server sends information to its destination via satellite and receives a response from the destination via satellite. Between a source and its destination, an in-bound channel and an out-bound channel are established to conduct communication. In the distance education, two channels are established between a main campus and a satellite campus. When there are two or more sites, the number of channels is increased, making this type of setup very expensive.

One-way transmission is widely used in distance education. In this system a transmitter is installed at the main campus, with receivers installed at satellite campuses. The lectures are broadcast from the main campus to the satellites simultaneously.

This system does not allow a question-and-answer session. However, Q&A between main campus and satellites can occur when they are connected via the Internet. Thus, the Q&A is conducted over the Internet and the lectures are broadcast via satellite. Figure 5.7 shows a diagram of the satellite/Internet system.

6

INFORMATION SUPERHIGHWAYS

The idea of a national information infratructure (NII) was proposed in the United States by the Clinton administration in 1993. Via NII, information such as images, pictures, videos, and speeches can be transmitted efficiently and effectively at a low cost. Through the introduction of NII, information can be exchanged freely and new business can evolve, for example, telecommuting, video on demand, virtual university, and teleshopping. In the past the construction of railways allowed such heavy industries as steel and ship building to evolve. Through the construction of highways, the automobile industry and express delivery services evolved.

The introduction of information superhighways can ensure the development of a new industry. Currently, television programs are broadcast to every home and viewers' choices are limited to what is already scheduled. However, when we have 156-Mbps subscriber lines, viewers will be able to receive video through the communication links in real time and access the TV station by phone to request a video program. This is what is known as *video on demand*. Similarly, we will be able to order 3D product information from a department store or request lectures to be transmitted from a university. We will access a travel agent for 3D interactive travel information. In this way teleshopping, tele-education, or teletravel can be accomplished. These applications are described in Chapter 11. In place of optical fibers for subscriber lines, asynchronous digital subscriber lines (ADSL) have been proposed in the United Kingdom and other European countries. This method uses a metallic cable to transmit information at 6–12 Mbps for downlink and at 640 Kbps for uplink. To enable this service ADSL modems are installed at home and at the telephone office. However, one drawback of ADSL is that the length of the subscriber line is limited to 1 or 2 km. Through the introduction of information superhighways, multimedia information can be transmitted in real time and new industries will be created.

For the next-generation Internet, two initiatives were announced. In 1996, the Internet 2 initiative was proposed by the university community of the United States. According to this plan, the backbone network of Internet 2 was to provide more than five times the transmission speed of the current backbone of the Internet, meaning 1- to 10-Mbps network service to the end users. It focused on security and quality of service (QOS).

The next-generation Internet initiative was announced by the Clinton administration. The initiative aimed to provide more than 100–1000 times the transmission speed of the current Internet. It focused on the real-time transmission of images or motion pictures. It was planning to be applied to telemedicine, telecollaboration, and/or tele-education. In 2000, the Japanese government announced a high-speed information infrastructure initiative by which high-speed subscriber lines would be constructed over the next five years to transmit motion pictures to subscribers in real time.

As information technology progressed, demands for enhancement of the network grew. For example, a high-speed network was needed to transmit a huge amount of multimedia information among institutes, including 3D images and motion pictures. To achieve this, projects to enhance the network were conducted in the 1990s. Two projects are described here: the gigabit network test bed project and the super-high-speed backbone network project.

6.1 THE GIGABIT NETWORK TEST BED PROJECT

The gigabit network test bed project was conducted by the National Research and Education Network (NREN) program until 1995. About 40 institutions, including universities, common carriers, computer manufacturers, and research institutes, participated in this project and had experiments for the development of application programs running on the gigabit network. There were five test beds in the project: Aurora, Blanca, Casa, Nector, and Vistanet. The project was inaugurated by the National Science Foundation (NSF) and the Defense Advanced Research Project Agency (DARPA).

The Corporation for National Research Initiative (NRI) was an organization for coordination among members of the consortium and the federal government. The government spent about $20 million for research over five years.

Common carriers and computer manufacturers donated the funds, which amounted to $400 million dollars.

As one of the research results in the gigabit network test bed project, the University of Illinois developed a CAVE, a virtual reality distributed environment. A viewer enters the CAVE's virtual space for an experience of virtual reality. Wearing special glasses with a sensor, he or she has a stereoscopic view of an object from different viewpoints. The system deforms the object and displays it to match the viewpoint.

6.2 SUPER-HIGH-SPEED BACKBONE NETWORK PROJECT

Among the initiatives for high-speed backbone network construction are the MCI project and the ACTS project, the latter conducted by NASA.

In the early 1990s, MCI, the long-distance common carrier, enhanced its backbone network from 155 Mbps to 622 Mbps. In 1998, MCI upgraded its network to 1.2 Gbps. The company proposed the very high-speed backbone network service (VBNS) in cooperation with the NSF. In 1993 the NSF decided to establish a 2.4-Gbps high-speed network for supercomputer centers in the United States and named it VBNS. A project on the modeling and visualization of weather forecasts was conducted using VBNS. The network has been maintained by MCI. Access to VBNS for the clients of supercomputing centers have been provided under this project. VBNS uses protocols such as Internet protocols and ATM protocols and provides 622-Mbps network services to end users.

The United State has conducted the projects on high-speed data transmission via satellites. One of them is the Advanced Communication Technology Satellite (ACTS) project, started in 1993, through which NASA has launched communication satellites. The project uses the KA band (30/20 GHz) and provides a bandwidth of 2.5 GHz. The transmission speed between satellite and base station was initially 156 Mbps, gradually increasing to 600 Mbps. ACTS has been used mainly for a distributed network of supercomputers. One application is the interconnection of NASA Goddard Space Flight Center in Washington, D.C., and the Jet Propulsion Laboratories (JPL) in California, which enabled an experiment on the simulation of hydrodynamics.

Under the ACTS project, the Keck telescope has been operated remotely from the California Institute of Technology and the data obtained from the telescope has been visualized. The network was established by linking mainland United States and Hawaii via ACTS and then extended by linking Hawaii and Japan via INTELSAT. The network has allowed analysis of data obtained from the telescope.

The ATM Research and Industrial Enterprise Study (ARIES), NASA, the Department of Energy, and common carriers have conducted joint research on probing for petroleum and on telemedicine, by interconnecting ACTS, terrestrial networks, and ships over the networks. ARIES has also conducted joint research on high-speed transmission by interconnecting the NASA Lewis Research Center and JPL via ACTS in cooperation with NASA and JPL.

6.3 INTERNET 2 AND THE NEXT-GENERATION INTERNET

The Internet has spread throughout the world rapidly. However, it has not provided high-speed transmission. This makes it very difficult to transmit multimedia information, such as images, motion pictures, and 3D images via the Internet. Demand is growing strongly to enhance the Internet.

In response, in 1996 two initiatives were announced by the United States to enhance the efficiency of the Internet substantially: Internet 2 and the Next-Generation Internet. The Internet 2 project was initiated by the university community of the United States. The Next-Generation Internet was initiated by the Clinton Administration.

6.3.1 Internet 2

Internet 2 is an intracampus high-speed network, called GigaPOP (gigabit capacity point of presence), providing a variety of Internet services. GigaPOP is linked to a high-speed backbone network BNS, which plays a role as collective entity (CE). In general, the transmission capacity of GigaPOP is about 622 Mbps.

6.3.2 Next-Generation Internet

The Next-Generation Internet (NGI) aims to be a speedier communication network, its goal being to provide 100–1000 times the transmission speed of conventional Internet access. Using NGI, real-time multimedia services such as a high-quality videoconference service are available. The number of users who have access to NGI has been increasing with the quality of services. For example, the current 64-Kbps Internet service has been enhanced to 6.4 Mbps or 64 Mbps. At 64 Mbps, motion pictures can easily be transmitted in real time. The 1.5-Mbps Internet service has been enhanced to 150 Mbps or 1.5 Gbps, enabling the large amount of information produced by a supercomputer to be transmitted in real time.

6.4 GLOBAL INFORMATION INFRASTRUCTURE

Following the Clinton Administration's 1993 announcement of the NII, the advanced countries, such as Japan, the United Kingdom, Germany, and France, announced their own information superhighway initiatives. Based on NII, in 1994 the United States proposed a global information infrastructure (GII) at the plenipotentiary meeting of the International Telecommunications Union (ITU). According to the GII initiative, anybody in any country can access GII and free competition can exist, GII can be enhanced without disturbing conventional service operations, and universal service can be provided through GII.

In 1994, at the Third International Conference on Broad Islands, held in Hamburg, Germany, the programs, perspectives, and activities and technological development concerning the Information Superhighways were talked about. One of the main events was an experiment on Information Superhighway transmission via interconnecting Berlin, Hamburg, and Cologne. A keynote address was broadcast from the headquarters of the Ministry of Telecommunications in Berlin to the conference venue in Hamburg. A joint experiment on Computer Supported Cooperative Work (CSCW), linking Hamburg and Berlin, was conducted via the German version of the Information Superhighway.

To achieve a global information infrastructure, the protocols and communication methods should be standardized. There are two methods for such standardization: a de facto standard, and an official standard, which is established by an international organization. But there are problems concerning standardization: If standardization, such as protocols and communication methods, is accelerated, then technological development could suffocate. Thus, coordination between the standardization activity and technological development is very important.

6.5 SIGNIFICANCE OF INFORMATION SUPERHIGHWAYS

In the 1960s, heavy industries such as shipbuilding and the steel industry evolved with the industrialization following the Second World War. In the 1970s, the electronics and machine industries (home electronics and automobiles) evolved. Progress in technology has both reduced costs and increased efficiency. To lower costs, manufacturing companies moved their factories to Asian countries, where wages were lower. But this strategy of exploiting the wage differential between advanced and developing countries has been less and less successful. To overcome this, a totally different strategy will be needed.

Progress in manufacturing industries has significantly increased petroleum consumption. It has become urgent to reduce petroleum consumption via the introduction of industries such as the information industry. Thus, information technology will play an important role in the twenty-first century, sometimes called the Information Age. Information Superhighways will be the information infrastructure in the Information Age.

7

NEWLY DEVELOPED TELECOMMUNICATION SERVICES

The union of information processing technology and telecommunications has led to the development and practical use of a variety of more human-friendly telecommunication services. For example, toll-free phone service has been widely used in advanced countries. Calls for flight reservations, hotel reservations, and online shopping are frequently toll-free. More than 50% of all subscriber calls are toll-free in the United States. In Japan, toll-free phone service is increasing year by year. A majority of calls throughout the world will be toll-free calls in the near future.

Caller ID display service has been introduced by means of ISDN network service. A subscriber can identify the caller's phone number before picking up a phone and can decide whether or not to pick up the call.

The mobile phone market is rising rapidly in many countries. In Japan, more than 60 million people use mobile phones. NTT DoCoMo, Japan, sells the i-mode mobile phone, which provides Internet access functions. It has been a great hit and has lead to a rapid increase in the number of Internet users in Japan.

We will now take a look at some of these new telecommunication services.

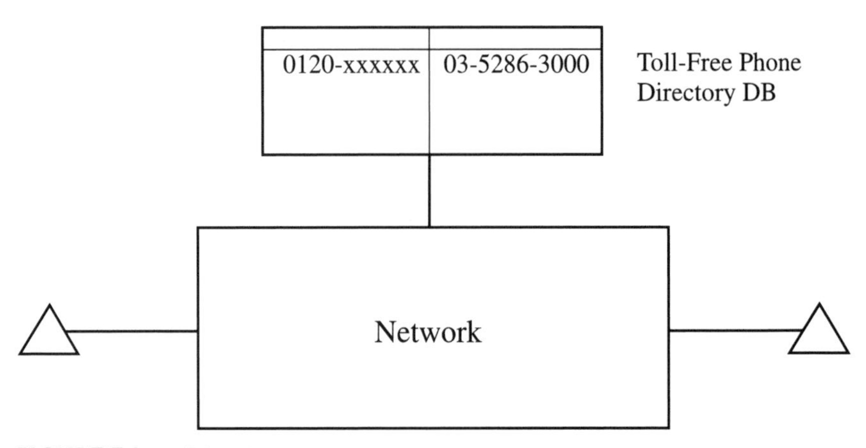

FIGURE 7.1 Toll-free phone system.

7.1 TOLL-FREE PHONE SERVICE

Toll-free phone service shifts the charges for a call to the destination, whereas telephone charges are usually made to the source. In Japan the prefix for a toll-free call is 0120; in the United States it is 800, 888, 877, or 866.

When a toll-free phone call is initiated, the system recognizes it as such, accesses the database of toll-free phone numbers, translates it into a destination number, and then switches the call to the destination (Figure 7.1).

The volume of toll-free phone calls is increasing every year. For example, in the United States the volume of toll-free calls has already exceeded that of the conventional calls. In other advanced countries, the number of toll-free phone calls is increasing rapidly.

7.2 CALLER ID SERVICE

The ISDN basic user network interface has two B channels and one D channel. Using the D channel, the caller's ID can be transmitted to the destination and displayed on the screen of the telephone. Thus, before answering the phone, the called party can see what number is calling, thereby allowing the called party to decide whether or not to answer the call.

Currently it is the caller's phone number that is displayed on the screen. This makes it difficult to identify the caller in a short time. It would be more desirable to show the caller's name on the display screen. Common carriers in Japan are currently investigating technology for translating a phone number into its subscriber's name in real time. To achieve this, a database consisting of a table of subscribers' names with their phone numbers has to be constructed and accessed in real time.

7.3 CALL FORWARDING SERVICE

Figure 9.12 presents an example of call forwarding service: Terminal A calls terminal B, which is busy, so the call is forwarded to terminal D. Terminal D is predesignated as the terminal to which a call is transferred should the called terminal be busy. Another service has a call transferred to a phone that has been predesignated as a forwarding terminal when a call is not responded to within a specified time. This might be useful, for example, at a small office when staff is away on business.

7.4 CALL WAITING SERVICE

Figure 9.11 shows an example of call waiting service: Terminal A calls terminal B, which is busy, connected to terminal C. So a signal is sent to terminal B informing it of the arrival of a second call. If the called party would like to take the call from terminal A, he or she clicks the phone, picks up the second call, and talks with caller A while caller C is put on hold.

7.5 MOBILE COMMUNICATION SERVICE

The mobile communication system serves mobile telephones, such as car telephones and portable telephones. Figure 7.2 shows the zone structure of the mobile communication system, where F1 to F7 show frequency assignment.

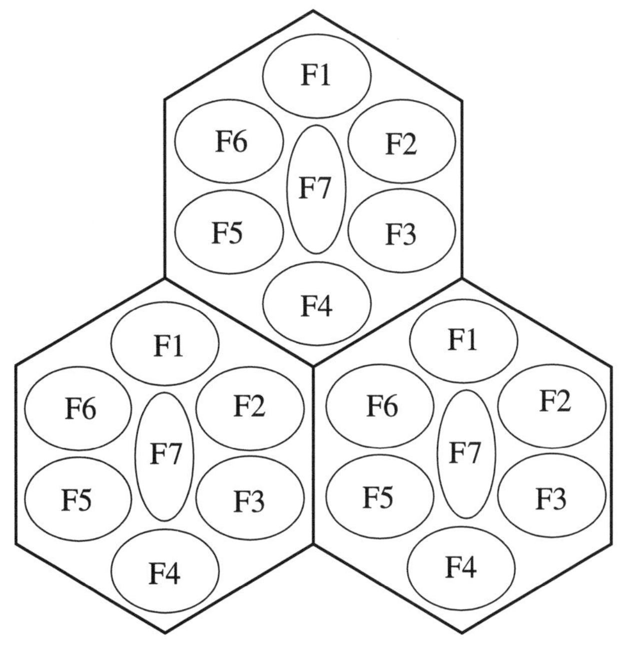

FIGURE 7.2 Mobile communication system zone structure.

F1 to F7 are cells, each with a diameter of about 10 km. Cells are laid out in hexagonal zones. Each zone has the same frequency structure. In this way, frequencies are used efficiently and effectively.

The mobile communication system comprises a mobile control station (MCS), a mobile base station (MBS), and a mobile service station (MSS). The system structure is shown in Figure 7.3. The MCS handles call control and management of the current position of the MSS. The MBS monitors the MSS and detects its current position. The MSS is a car telephone or a portable phone that issues its identification number and lets the MBS know its current position. The MCS has a home memory and stores the current position of the portable phone.

When an MSS is switched on, the identification number (ID) of the MSS is issued. An MBS that captured the ID sends it to the MCS controlling the MBS. The MCS stores in its home memory the information about the MSS and the area of the MBS where the MSS is currently located.

7.5.1 Calling a Portable Phone from a Wired Phone

A wired telephone calls a portable phone. The call is identified as a portable phone call by the dialed number. The numbering plan for portable phones is as follows:

prefix (3 digits) + carrier ID (3 digits) + subscriber number (5 digits)

(The mobile prefix number in Japan is 090.) The call is transmitted via the gateway to the MCS, which checks the MSS current location and transmits the call to the MBS closest to that location. Then the MBS calls the MSS and the call is connected to the MSS.

7.5.2 Calling a Portable Phone from Another Portable Phone

A portable phone calls another portable phone. The call is identified as a portable phone call by the dialed number. The call is transmitted to the MCS, which checks the current location of the MSS and transmits the call to the MBS closest to that location. Then the MBS calls the MSS and the call is connected to the MSS.

7.5.3 Calling a Wired Phone from a Portable Phone

A portable phone calls a wired phone. An MBS captures the call and identifies it as a call to a wired phone. The call is sent to the telephone network through the MCS and gateway. Then the call is connected to the wired phone.

7.5.4 I-Mode Phone

In 1999, NTT DoCoMo, Japan, announced it would sell the i-mode mobile phone in Japan. It was an epoch-making announcement because the phone could access the Internet. Using the i-mode phone, a customer can access the Internet,

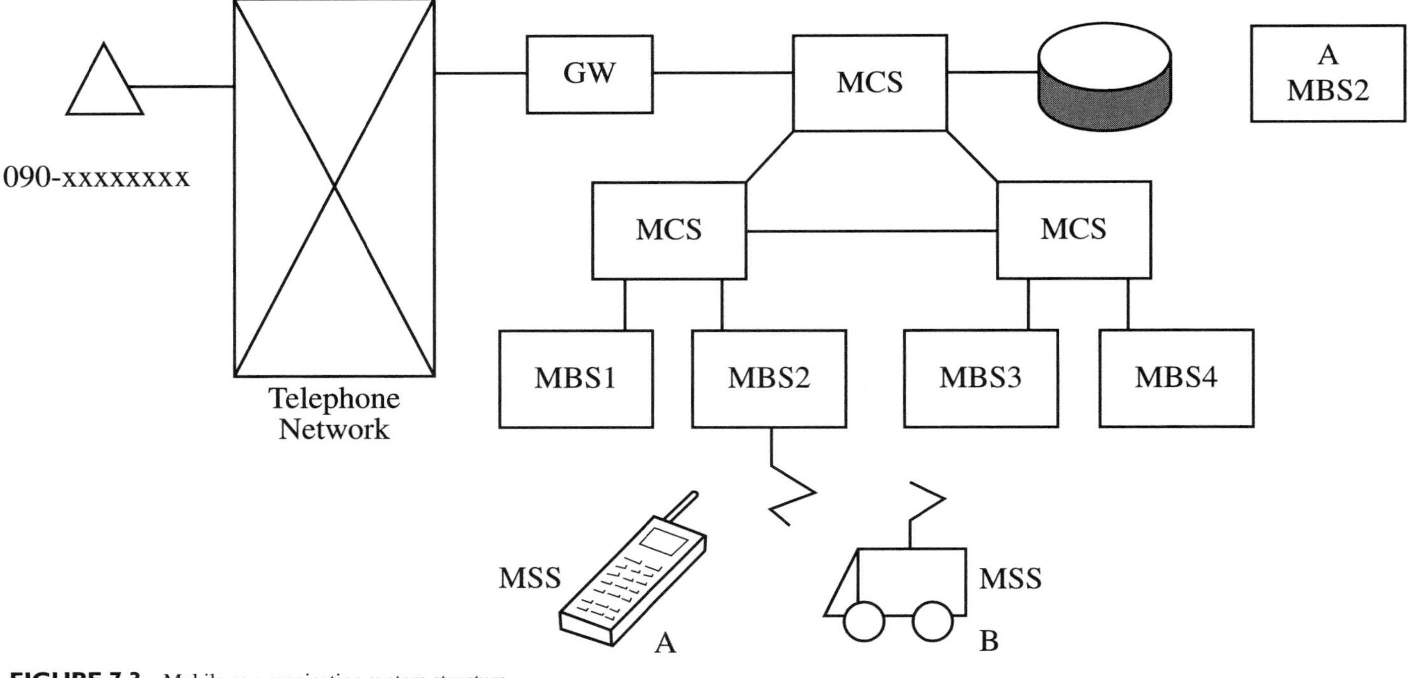

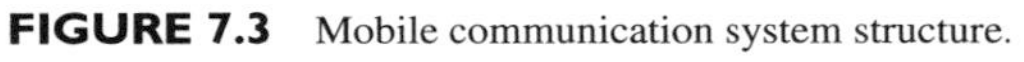
FIGURE 7.3 Mobile communication system structure.

get information, and exchange e-mail messages. It was a great hit in the mobile phone market in Japan. By the end of 2000, the number of i-mode phones sold was about 17 million.

In January 2000, NTT DoCoMo, Japan, began to sell more advanced i-mode mobile phones in which the Java program environment was installed, in cooperation with Sun Microsystems. Java programs in the server are downloaded to a mobile phone via Hypertext Transfer Protocol (HTTP). Communication between a mobile phone and a server is conducted by means of HTTP. There are three application programs written in Java: a stand-alone application program that runs only on a mobile phone, a client–server application program that serves communication functions between client and server, and an agent-type application program that starts to run automatically every specified time, communicating with the server and getting information from it. In the mobile phone, a Secure Sockets Layer has been installed for security. In the i-mode mobile phone, all kinds of application programs, such as entertainment programs, can be downloaded from the server and can be executed. The i-mode mobile phone can provide communication functions, act as a station for Internet access, and execute a Java program.

7.5.5 Personal Digital Assistants

According to the report of the NPD Group, USA, shipment of personal digital assistants (PDAs) is increasing rapidly. In 2000, the number shipped reached 3.5 million units in the United States. The PDA can be used in stand-alone mode but also in communication mode. Palm, Inc., and Handspring, Inc., are frontrunners in the PDA market. Hewlett-Packard, Compac, Casio, and Sony have entered the PDA market as well.

7.6 THE INTERNET

The Internet is the network of networks. That is, local area networks are interconnected by telephone lines and/or dedicated lines. The basic component of the Internet is a local area network. In this section the network components, network topology, and media access control method of a local area network are described. Then the Internet and Intranet are described.

7.6.1 Local Area Network

Local area networks are widely used in the intranets in offices, colleges, universities, and factories. For example, a local area network is constructed in an area covering 5–6 kilometers in a factory by linking large-scale computers, workstations, personal computers, file servers, and control processors. All of the computers and terminals are linked, which makes them more efficient and effective than in stand-alone use. Linking the equipment enables resource sharing and functional distribution.

The main characteristics of the local area network are as follows.

(1) Coverage of the network is within a room, a floor, a building, or an area up to 10 kilometers.
(2) Transmission speed is over 1 kbs.
(3) The number of nodes connecting to the network is about 10–1000.
(4) Network topology is rather simple and the network has high flexibility and reliability.
(5) Communication control is simple and inexpensive. The network can be expanded by linking to additional networks via gateways or routers.
(6) All kinds of terminals or computers can be accommodated in the network.

An example of the local area network is shown in Figure 7.4.

Supercomputers, database machines, file servers, terminals, and workstations are linked together by a local area network. By accommodating a variety of workstations in the local area network, functions such as file transfer, message handling, logging-in to a remote terminal by a virtual terminal, and/or distributed file systems using network file system (NFS) can be achieved.

Through the evolution of workstations, personal computers, and local area networks, users have come to prefer the small computers, such as personal computers, to large-scale computers. This has contributed to the downsizing of computers.

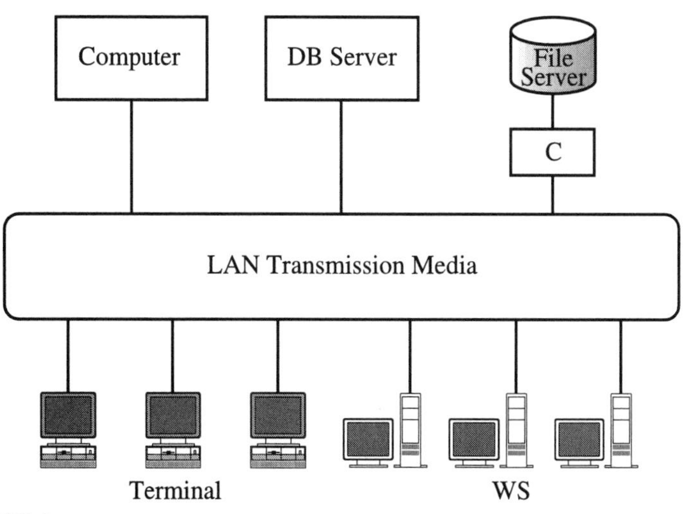

FIGURE 7.4 Local area network.

Distributed computing and the prevalence of small computers has reduced the cost of installation, operation, and maintenance of the LAN.

Each LAN is treated as a component of the global network. A number of local area networks are interconnected in metropolitan networks. In this way the network structure is expanded. And through this, a user in the local area network can communicate with any users in another local area network via the Internet. The Internet will be the network infrastructure for everyone in the world.

7.6.2 LAN Topology

In a local area network, coaxial cables, optical fibers, or twisted-pair cables are mainly used as transmission media. The star, bus, and/or ring topologies are used as the LAN topology. These three types of LAN topologies are shown in Figure 7.5.

In the *star* network, each node is linked to the center node. This network is a typical one in the global network. The center node is common and performs the switching functions for all. Each node is linked to the center via its transmission line. The star topology is used mainly in the intraoffice network. Twisted-pair cables or optical-fiber cables are used as the transmission lines. Its merits are that it is easy to manage and appropriate for peer-to-peer communications. A disadvantage is that it is not reliable, expandable, or flexible, because the center node performs all of the switching functions—once it goes wrong, the network is in trouble.

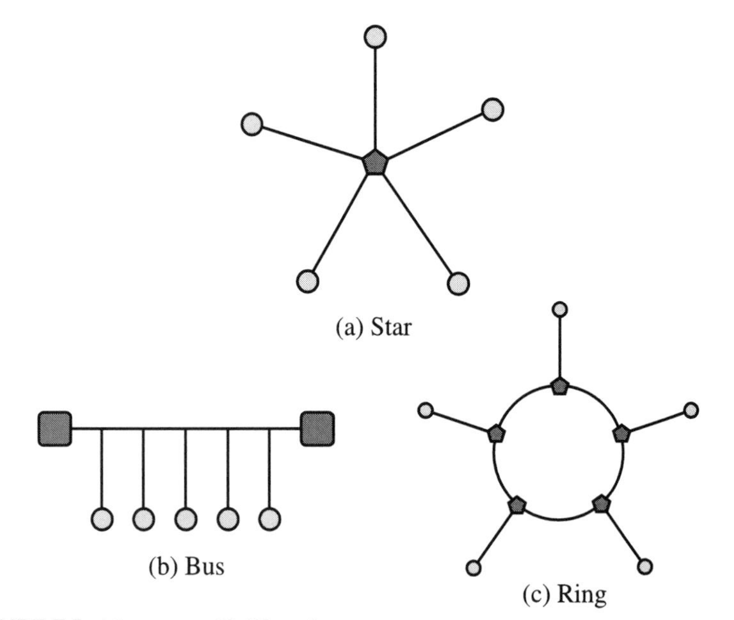

FIGURE 7.5 Three types of LAN topology.

In the *bus* network, each node is linked to a bus that is common to all nodes. The information issued by a node is transmitted via the bus. Coaxial cable is mainly used for the bus network. Every node has its communication function. Even when a node is in trouble, it produces no negative effects on other nodes. Therefore the network is reliable. And the network structure is easily changed and expanded. Any node can be linked to the bus at any time. Many channels can be accommodated in this network by supplying the different carrier frequencies to make it a broadband network. However, the network is low on the security of the data transmitted among nodes. Finally, when the bus is in trouble, all of the traffic goes wrong.

In the *ring* network, twisted-pair cables or optical-fiber cables are used. One-way transmission is allowed in this system. The network is a closed one, and each node is in sequence. When a node in the ring issues a message into the network, it is transmitted via the ring until the message returns to the original node. The message becomes distorted while running through the ring. But each node that receives it reproduces and passes it to the adjacent node. Therefore the distance of the ring can be expanded. The bus and ring networks act as multicast transmission systems. That is, any node in the network can issue a message and receive it. By introducing adequate access control technology, some kinds of switching and broadcasting functions can be achieved.

In addition to the aforementioned LAN topologies, there are modified LAN topologies, such as the *U bus* network. When coaxial cables are used for a broadband network system, it is necessary to install head-end equipment to act as signal amplifiers. The U-type bus system is shown in Figure 7.6. The signals are transmitted from the head end to the client via the cable. The cable is called an outbound line and is used only for the transmission of the input signals. On the other hand, the signals are transmitted from clients to the head end by the cable. The cable is called an inbound line and is used only for the transmission of the output signals. To achieve inbound and outbound transmission over a single cable, a frequency

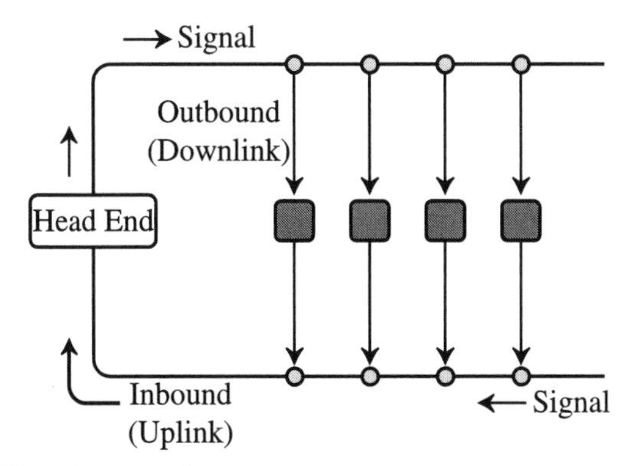

FIGURE 7.6 U-type bus network.

division multiplication method is used to make communication efficient and effective. This type of topology is used for the metropolitan two-way cable television system.

To construct a bus network with one-way transmission lines such as optical cables, the *star bus* topology is used. In the center of the network, a star coupler is installed. Each node in the network has a transmitter and a receiver. The star coupler has input terminals and output terminals. The transmitter of each node is linked to an input terminal, and the receiver is linked to an output terminal. The signal issued by a node is sent to all of the nodes via the star coupler. This behaves just as in the bus network.

There is another ring network: Two wires are linked to the center node, which is called a *concentrator*. This network topology composes a ring network via the concentrator. When a node is in trouble, it is bypassed; communication can be performed except for the troubled node.

Recently, wireless local area networks have evolved. Using frequencies up to 10 Mbps, communication is conducted between a mobile terminal and a base station positioned at the ceiling. It is used for in-house communication. The network topology is a star network that provides 1 to *n* communication channels.

7.6.3 Media Access Control Method

The transmission media in a bus or ring network is used as a multicasting transmission line accessed by every node. A source node deblocks a message into several blocks, adds the source address and its destination address to each block of the message to compose an information block, and sends the information block to the transmission line with time-division mode. The information block is called a *frame*. Each node in the network receives a frame and accepts the frame that is designated to the node while deleting the other frames. Through this action, each node exchanges information.

For each node to share the same transmission media in a consistent way, an access control method is needed. This is called the *media access control*. The fundamental control principle is equivalent to the media access control method of packet switching networks. An access control method is one of the main features of the local area network. The typical methods are described next.

7.6.4 CSMA/CD

CSMA/CD is a typical protocol for the Ethernet and is a simple and effective communication method in a local area network. In this method, any source node issues one frame at a time. When only one node issues a frame at a particular time, the frame is sent to its destination successfully. However, when two or more source nodes issue their frames at the same time, a collision occurs and the frames are destroyed. The nodes that issued their frames at the same time back off and reissue their frames at random times until transmission succeeds.

CSMA/CD is carried out in the following sequence.

(1) *Sending a frame:* Using a carrier detection technique, each node monitors the transmission medium. If a carrier is detected and another node is sending a frame, the first node must wait until the transmission medium is idle. When the transmission medium becomes idle, any node can issue its frame into the medium after waiting for the time slot in which a minimum frame is transmitted. If no collision occurs during the transmission, the node finishes the sending operation and waits for the next request to send a frame.

(2) *Collision detection and recovery:* When a node issues a frame, it tries to detect a collision. When a collision is detected, it immediately stops sending, issues a jam signal to the transmission medium, and informs it of the occurrence of the collision. After stopping its operation, it backs off and waits a while before trying to reissue the frame. The waiting time is determined by the *N*th power of 2, where *N* is the number of the collision. This is called the *binary exponential back-off* method.

(3) *Recovering a frame:* The receiving operation starts with the detection of a start frame delimiter by getting the preamble of the frame. The operation ends after checking the frame address, frame length, and frame check sequence (FCS).

CSMA/CD is a well-known method that gives efficient and effective transmission when traffic is scarce. That is, when the number of source nodes is small, high-speed transmission is possible because full use is made of the transmission network.

When traffic is heavy, that is, when it occupies more than 70–80% of transmission capacity, performance decreases substantially because of collision of the frames. This makes it is difficult to predict the completion time for the frame's transmission. This defect of CSMA/CD makes it inappropriate for applications such as production management and real-time control of a factory. However, it is good for the intraoffice environment where requirements are not severe.

7.6.5 Token Ring Protocol

In the token ring method, a token is circulated through the network. A source node gets the right to transmit a frame once it acquires the token. A node that has a frame to transmit waits to acquire the token, at which time it begins to transmit the frame. It releases the token back into the network after finishing the transmission. As the right to transmit a frame is moved from one node to another in the token ring, no conflicts among nodes can occur. Therefore, performance is very high when traffic is heavy in the network. The token ring method is divided into two submethods depending on the network topology.

(1) Token ring method: This is applied to the ring network.
(2) Token bus method: This is applied to the bus network. Since a bus network is not a ring topology, a logical ring is constructed and a token is circulated in the logical ring network.

Once a source node acquires a *free* token, it changes the token into a *busy* token and issues a frame following the busy token. Fundamentally, the transmission of a frame is permitted with a busy token. When the transmission is successfully performed, the source node issues a free token into a ring network.

In the *single-token* method, the node that acquired a busy token issues a frame following the busy token and finally receives the busy token, which runs through the ring network. The transmission of the frame is defined as finished. The time when the node receives a busy token is defined as the time of completion for transmission of a frame. In this method, only one free token or busy token is running through the ring network.

Figure 7.7 presents an example of the single-token method. In Figure 7.7(a), a free token is running through the ring network. In Figure 7.7(b), node E acquires a free token and issues a frame following a busy token. In Figure 7.7(c), node B receives the frame and releases the frame with the busy token. In Figure 7.7(d), node E receives the frame with the busy token and releases a free token into the network. As said before, in this method only one token is running in the network.

In the *multitoken* method, the time when a node that acquires a free token finishes issuing a frame is defined as the time of completion for transmission of the frame. When the transmission is complete, the node releases a free token into the network. In this method, a busy token and a free token exist at the same time in the ring network. Generally speaking, one free token and two or more busy tokens exist in a ring network.

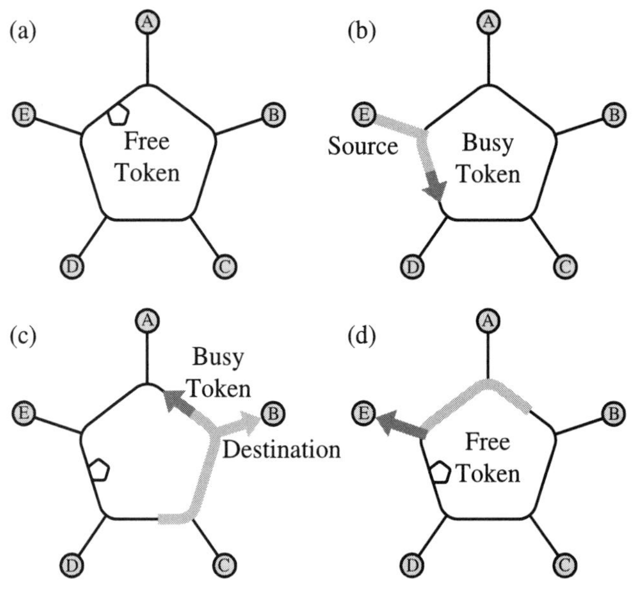

FIGURE 7.7 Single-token method.

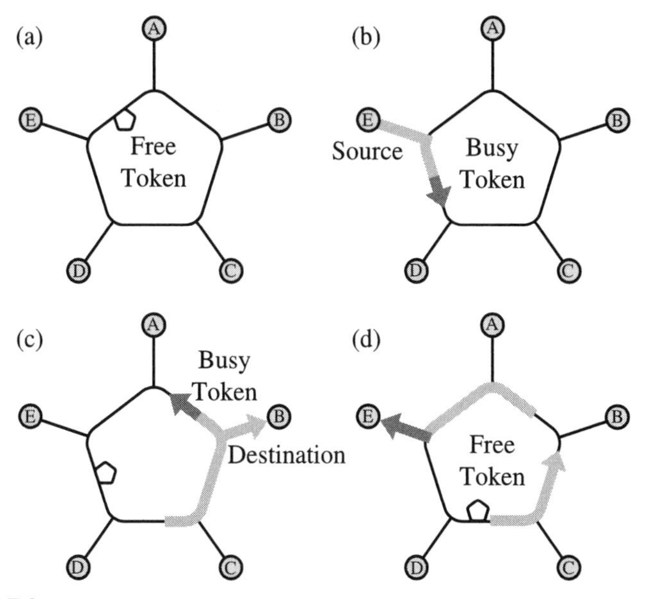

FIGURE 7.8 Multitoken method.

An example of multitoken method is shown in Figure 7.8. In Figure 7.8(a), node E acquires a free token. In Figure 7.8(b), node E issues a frame following a busy token and then releases a free token. In Figure 7.8(c), node B receives the frame issued by node E following the busy token. In Figure 7.8(d), node D acquires a free token and issues a frame following a busy token. After issuing a frame, it releases a free token.

In the example of Figure 7.8, two frames and one free token exist in the ring network. In the single-token method, frame transmission performance is relatively low. Because only one token exists in the network, it is easy to manage. This method has been standardized as IEEE 802.5 token-ring method. In IEEE 802.5, while a free token is processed bit by bit by using a shift register, the free token is changed to a busy one, after which a frame is issued. This method is called on-the-fly processing.

When a ring network is long or the transmission speed is high, the ratio of frame transmission time to a cycle time becomes small. In this case, transmission performance worsens. To illustrate this, consider the transmission of an 800-bit-long frame in the following two cases.

- Case 1: The ring network is 2 km in length and the transmission speed is 4 Mbps.
- Case 2: The ring network is 200 km in length and the transmission speed is 4 Mbps.

In Case 1, cycle time is about 10 microseconds. In case 2, cycle time is about 1,000 microseconds. Transmission time of a frame is 200 microseconds in both Case 1 and Case 2, so in Case 1 the transmission time of a frame is 20 times the cycle time and the transmission performance is not affected in the single-token method. In Case 2, the transmission time of a frame is one-fifth the cycle time. Twenty percent of the transmission capacity is used and 80% is unused.

7.6.6 Token-Bus Method

A source node issues a message both ways through a bus network. Because a bus network is not a ring network, it is difficult to circulate a message in the network. Therefore a logical ring network is constructed virtually on the bus network. Each node is assigned a sequential number. A token is transmitted from one node to the adjacent node. A token is a control frame composed of a source address and a destination address. A node that has a request to send a message waits for the arrival of the token that is designated to the node. Once the node receives the token, it is possible to send a message during the allotted time. When it finishes sending a message or the allotted time has passed, it issues a token that is designated to its adjacent node. And it waits for the arrival of a new token. In the token-bus method, priority control of transmission is performed by the time token control method. When traffic is light, token cycle time is short; when traffic is heavy, token cycle time is long. When a token circulates at high speed, the network has light traffic, so the possibility is great that a low-priority node can issue a message. When a token circulates at low speed, the network has heavy traffic. Then a low-priority node has little chance to issue a message.

To summarize, here are the main characteristics of the token-bus method and the token-ring method.

(1) In the token-ring method, a token has an identifier whose size is 8 bits. But because a prefix of 8 bits and a postfix of 8 bits are attached to the token, the size of a token is 24 bits. In the token-bus method, a token consists of a token identifier, a source address, and a destination address.

(2) In the token-bus method, the source node that issued the message monitors the completion of transmission of the message to the next node.

(3) In the token-bus method, the network configuration changes with the addition of a new node, the removal of a node, or the malfunction of a node.

(4) In the token-ring method, the network is closed, and it is necessary to remove the information that is circulated through the network.

(5) In the token-bus method, the network is passive. Therefore it is reliable in comparison to the ring network. The token-bus method has the characteristics not only of a bus network but also of a logical ring network. The token-bus method has been standardized as IEEE 802.4.

7.6.7 Slotted Synchronizing Method

In the slotted synchronizing method, a transmission medium is partitioned into a slot the size of a frame or a unit smaller than a frame. A message is transmitted by using the slot assigned to the source node. This is called the TDMA method. In TDMA, a slot is assigned to each node. TDMA gives each node an equal opportunity to use the transmission medium and stable delay time characteristics. However, it is not appropriate for a burst of traffic because a slot is assigned to each node whether it is used frequently or not. Another method that has been proved to be efficient, based on TDMA, is the slotted-ring method. In this method, a message is transmitted through the ring network using any time slot in a slotted ring. When a source node sends a message, it uses a time slot for sending and releases it after finishing a transmission. In this case a source node occupies the time slot when it has a message to send. Therefore network efficiency is substantially high.

7.6.8 Local Area Network Standardization

Local area network standardization has been conducted mainly at the Institute of Electrical and Electronics Engineers (IEEE) 802 committee. The International Standards Organization (ISO) and the American National Standards Institute (ANSI) have also studied standardization issues. The methods that have been standardized so far are CSMA/CD (as IEEE 802.3 and IS 8802/3) and the token-bus method (as IEEE 802.4 and IS 8802/4). FDDI has been standardized by ANSI.

7.6.9 Layered Structure of a Local Area Network

A local area network protocol has a layered structure and corresponds to the OSI reference model. The correspondence between the OSI reference model and an LAN protocol is shown in Figure 2.4. The data link layer protocol corresponds to the Logical Link Control (LLC) and Multiaccess Control (MAC) protocols. The physical layer protocol of the OSI reference model corresponds to the physical layer protocol of the LAN. The LLC protocol provides a common data transmission function to the upper layer of the LLC. The LLC frame consists of a data service access point (DSAP), a sender service access point (SSAP), control information, and data. LLC has been standardized as IEEE 802.2 LLC. In MAC, there are CSMA/CD, token-bus, token-ring, and FDDI MAC protocols. CSMA/CD, token-bus, token-ring, and FDDI MAC protocols have been standardized as IEEE 802.3, IEEE 802.4, IEEE 802.5, and FDDI (ANSI) protocol, respectively. In CSMA/CD, there are 10 BASE 5, 10 BASE 2, 10 BASE T, and 10 BROAD 36 protocols. FDDI has been standardized by the American National Standards Institute (ANSI) and is used mainly for a high-speed LAN. The coverage of the network based on FDDI is about 40–50 km.

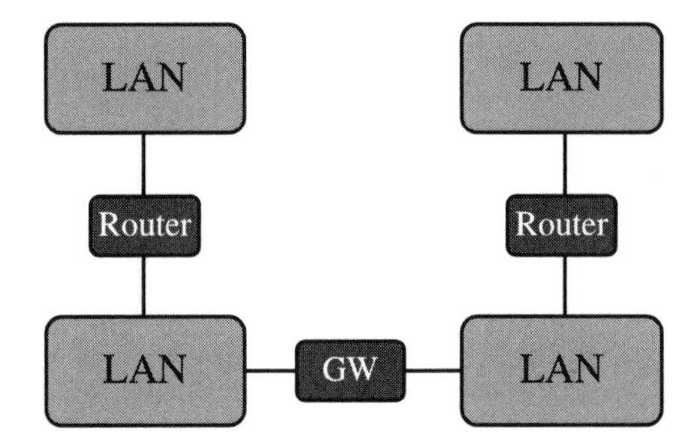

FIGURE 7.9 Internet topology.

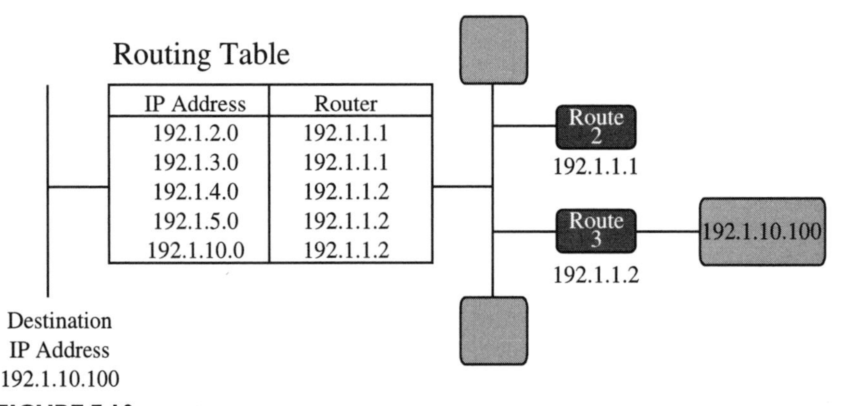

FIGURE 7.10 Routing.

How does a local area network work? An example involving the Internet is shown in Figure 7.9. In this network, LANs are interconnected by routers and gateways (GW).

(1) *Function of the router:* The router sends a message to the router designated by the routing table. It does not perform protocol conversion.

(2) *Functions of a GW:* In addition to routing, a GW performs protocol conversion.

(3) *Example of routing:* An example of routing is shown in Figure 7.10, in which the designation IP address is 192.1.10.0 and the corresponding router address 192.1.1.2 is chosen. According to the information, router 3 is selected and the message is sent to router 3.

(4) *TCP/IP protocol example:* A TCP/IP protocol example and a message form are shown in Figure 7.11. The destination IP address is 1.11.3, which identifies the destination node to which the message is sent. The destination port number is 2001, which identifies the destination process to which the message is sent.

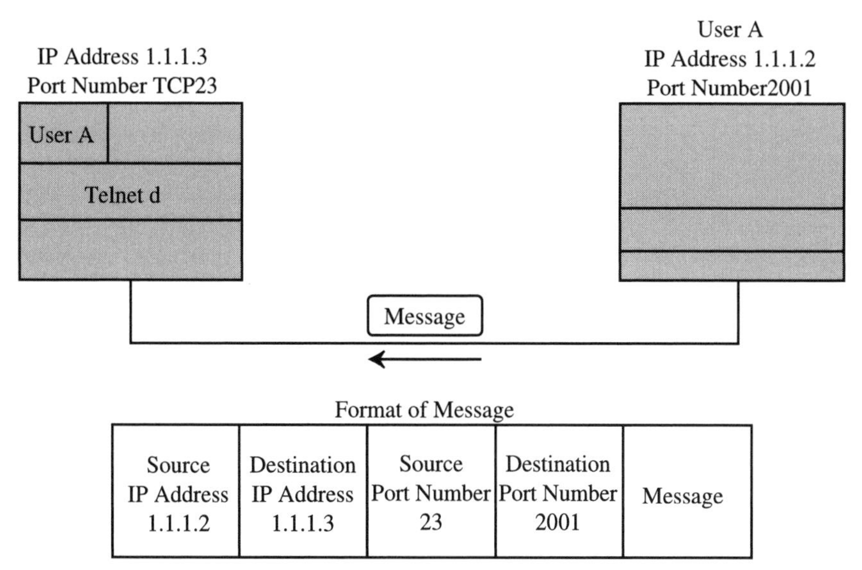

FIGURE 7.11 TCP/IP protocol example.

7.7 INTRANET

An in-house equivalent of the Internet is called an *intranet*. Intranets can be connected to the Internet to give a user access to resources such as files or databases on the Internet. On the other hand, this means that the company's confidential information may be accessed from outside, via the Internet. To protect such confidential information, a firewall is constructed. Information that is not confidential can be accessed from outside. An example of an intranet is shown in Figure 7.12. Both the Internet and an intranet are local area networks interconnected by routers or gateways.

7.7.1 Firewalls

A firewall is used to protect an intranet from the attacks of hackers or dishonest users. If there is no firewall at the entrance of an intranet, anyone may enter it and steal secret or private information. A firewall is a system to protect information. The firewall can be installed at various points of the intranet. It can be installed between LANs to protect the information of a department or section of the company. It can provide not only filtering but also encryption, authentication, and virus detection.

A basic function of a firewall is IP filtering. Because an IP address is used for routing of TCP/IP, the IP address is used to check a source and a destination. The firewall knows which application is used by checking a port number. By means of a three-way handshake, it knows which is the source and which the destination.

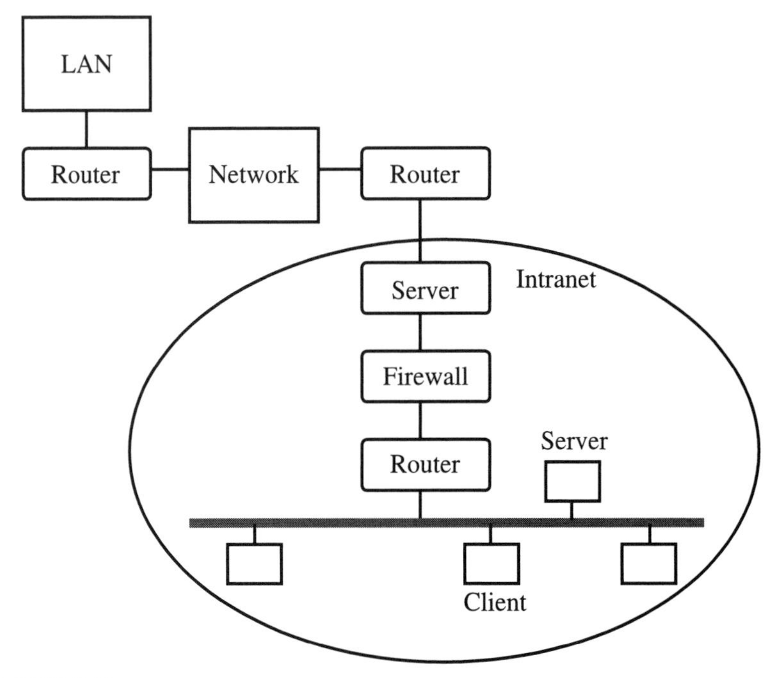

FIGURE 7.12 Intranet example.

An IP filter provides functions of packet control using this information. It checks which packet is allowed to pass through the firewall. Usually a packet issued from inside the intranet is allowed to pass through the net, but a packet from outside will be checked carefully to see whether it should be allowed to pass or not. When IP filtering rules are implemented, the router picks up a rule and compares the header of a packet with it. When the rule and the header are coincident, the packet is denied passage through the firewall. If they are not coincident, the router picks up the next rule and compares the header with it. When they are coincident, the packet is rejected. This procedure continues until the end of the rules.

7.7.2 Intranet Protocols

There are two intranet protocols: CSMA/CD and token ring. These are described earlier in the chapter. This section describes the features.

7.7.2.1 CSMA/CD

This topology was invented by Xerox Corp. It is a local area network connected by a bus interface named Carrier Sense Multiple Access/Collision Detection (CSMA/CD). Here's how this method works: Any workstation on the intranet may issue messages at any time, meaning that more than one message can be issued at the same instant. If messages are transmitted through the bus at the same time and a conflict occurs, then the source waits a moment and then issues the message again. When the message is reissued it is determined by, for example, using a table of random numbers. This method is similar to a roundtable discussion, where someone begins to talk and other people wait until the first speaker finishes before they talk. When more than one person begins to talk at the same time (a conflict), one speaker stops talking and waits until the other finishes. In this way the discussion can continue.

7.7.2.2 Token Ring

In this method, a token goes around the ring circuit. Terminals are accommodated on the ring. When a terminal is ready to send a message to another terminal, it catches the token, places a message with a destination ID in the token, and then releases the token on the ring circuit. The destination terminal catches the token, reads the message, writes an acknowledgment on the token, and releases it on the ring circuit. In this way, message passing can be performed smoothly and without conflict.

7.8 CONTINUOUS ACQUISITION AND LIFELONG SUPPORT

Although continuous acquisition and lifelong support has been acronymed CALS, more recently CALS has come to mean "commerce at light speed"; that is, CALS provides the means, tools, or systems for conducting business transactions at light speed. In the business world, there are a variety of documents, such as a written order, a document for maintenance, and/or a document for sales. Each has a different form. It is therefore necessary to access each one separately if we want to know specific information. This takes time. Generally speaking, documents have different styles, are written in different languages, and are made by different processors. The number of documents has been increasing year by year. To reduce the number of documents and to access the information efficiently, CALS was proposed. In this system, written orders or documents for maintenance or sales are integrated into a single document, and all of the information can be accessed by reading the integrated document. In CALS, the style of the document is standardized, and it can be used during the life cycle of goods, namely, order, development, sales, and maintenance.

7.8.1 Features of CALS

(1) The document is digitized, enabling efficient processing.

(2) The document is standardized; that is, the style and the form are defined.

(3) CALS is an open system and runs on any kinds of hardware or operating systems. Thus it enables data interchange to be performed efficiently and effectively.

Personnel of the design, manufacturing, maintenance, and sales departments can access the document in standard form, use it for given purposes, write information in the document, and refer it to other personnel.

7.8.2 Example of CALS

(1) *Standard for documents:* Standard Generalized Markup Language (SGML) is the standard for the document. Tags are put on the beginning and end of the document, the chapter, and the section. This allows easy retrieval of the document by chapter or section.

(2) *Standard for figures:* Computer Graphics Metafile (CGM) is the style standard for figures and illustrations.

(3) *Standard for CAD:* Initial Graphics Exchange Specification (IGES) is the standard for CAD (computer-aided design).

7.8.3 Details of CALS

Because of information sharing following the introduction of CALS, the procurement of components and products has been expanding on a global scale. In recent years, companies have moved their factories and their research and development institutes to various parts of the world. Product distribution and inventory management have to be conducted throughout the world. To do so efficiently and effectively, companies must introduce CALS. Information shared via CALS concerning the manufacturing process or procurement can easily be processed. This information helps companies devise strategies and make plans for the development of new products.

7.8.4 Features of CALS

Sharing information in digital form is one of the important features in CALS. The major objective of CALS is to increase the efficiency of the manufacturing process. Therefore, CALS can be applied to a variety of applications, such as concurrent engineering and virtual corporation. Sometimes it is said that CALS is electronic data interchange (EDI). But EDI is actually the specification of documents used in commercial transaction, and is only a component of CALS.

7.8.5 Components of CALS

There are three components of CALS.

(1) *Standards for information interchange:* Items for standardization are the data formats, such as documents, figures, product models, voice and motion pictures,

definitions of data, description methods, sequences of transmission, and the layouts of data components.

(2) *Rules for information interchange and sharing:* Items are specified, which include the time, frequency, and priority of information interchange, the certification of information interchange, the acknowledgment of information, security, and database specifications.

(3) *Related software:* There is a software that converts documents written by means of word processors to those specified by SGML.

7.8.6 Prerequisites for Information Sharing

Three elements of CALS have to be considered to establish information sharing: the digitization of information, standardization, and the open system approach. Generally speaking, data formats, specifications, communication methods, and protocols differ among companies or institutions. In these situations, it is difficult to share information among the different organizations. But by using CALS, different companies can share information by interconnecting via communication networks.

To share information, it is appropriate to digitize the information, including not only text but also figures, images, voice, and motion pictures. Information has to be digitized in the process of manufacturing, including design, development, manufacturing, sales, and maintenance. Conventional digitization of information had not been standardized, but was done by using the computers or systems that system integrators had. Digitization depended on the specific computers or systems, so it was difficult to exchange information. Via the introduction of CALS, the formats for information and communication protocols have been standardized. By taking the three elements—digitization, standardization, and the open system approach—into consideration, it is easy to exchange information among different companies.

In order to implement CALS, it is important to construct an integrated database (IDB) that serves the functions of information retrieval in digital form. IDB is a distributed database linked over the network. By means of IDB, information that is distributed over the network can be shared.

7.8.7 Standardization of CALS

The standards that have already been established are introduced as the elements of CALS. The specifications that have been standardized by international organizations of standardization, such as ISO, are studied first. Second, the specifications that have been standardized by domestic standardization organizations, such as ANSI, are studied. Third, the specifications that have been used widely in the private sector as de facto standards are studied. If there are no models for standardization of CALS, the company studies its own specifications for standards.

In summary, the company does not begin from scratch to make specifications as the standards for CALS. They study the specifications that have already been standardized and then judge whether they are suitable or not for the standardization of CALS.

A struggle for leadership occurs when standardization activities proceed. In the case of CALS, the EU and the United States competed for leadership. For example, for the standardization of EDI, the United States claimed ANSI.X12, whereas the EU claimed EDIFACT. But cooperation among competitors is most important in order to coordinate and establish standards.

There are two ways to handle standardization. One is to take into consideration de facto standards, that is, the specifications that have been widely used in the private sector, such as the Windows operating system developed by Microsoft. The other approach involves the official standards adopted by the standardization organizations, such as the ISO standards.

The open-system approach has been proposed because that enables different types of computers to interconnect and makes the exchange of information possible. Not only the standardization of data formats but also the standardization of the information interchange have been taken into consideration in CALS. By the establishment of the standardization of the data formats and data interchange, data can be exchanged freely among the different systems. CALS assures the information interchange and interoperability among systems.

7.8.8 Organizations of Standardization

There are many organizations for standardization, from the global level to the domestic level. Here are the main organizations. The International Standards Organization (ISO), established in 1947, is the most famous. Its major focus is standardization of items related to industry, science, and technology, except for telecommunications issues. Only one organization from each country is allowed to participate as a member. The International Electrotechnical Commission (IEC) was established in 1906. Its task is to standardize items related to electrical and electronic subjects. For the standardization of computers, ISO is in charge of computer software and IEC is in charge of computer hardware.

In 1987, a joint committee of ISO/IEC, JTCI, was organized to handle the cooperative standardization of computers. The International Telecommunications Union (ITU) is the professional organization for standardization of telecommunications under the umbrella of the United Nations. The American National Standards Institute (ANSI) is the ISO member representing the United States.

The Institute of Electrical and Electronics Engineers (IEEE) is the organization for standardization of computer systems and information processing in the United States. About 600 standards are considered every year for revision. The Computer and Business Equipment Manufacturers' Association (CBEMA) is in charge of the standardization of computers and business machines, under the umbrella of ANSI.

7.8.9 CALS Standards

The CALS standards that have been standardized by ISO are SGML, CGM, IGES, and STEP. Some of their important specifications are described next.

7.8.9.1 Standard Generalized Markup Language (SGML)

Standard Generalized Markup Language (SGML) is the standard for describing text. Generally speaking, the markup is used to edit a document. It specifies the beginning and end of the text, the writing styles, and the fonts. The markup specification depends on the manufacturer and therefore may be incompatible with another specification. For example, it is possible to exchange the textual data of a document created by MS-DOS. However, it is difficult to exchange gothic characters and tables written by means of MS-DOS. With the introduction of SGML, it became possible to exchange a document that includes gothic characters and tables. One of the major functions of SGML is to create a document using the skeleton of the document. SGML provides a document-type definition (DTD) facility by which a document can be constructed in structural form. For example, if a document consists of a title, an abstract, chapters with sections, and a conclusion, SGML functions can put the document in its structural form. This also makes it easy to revise the document by erasing a part of it or adding a new part. It is easy to handle the document written by SGML in a database, because the document is structured into, say, its title, abstract, chapters and sections, and conclusions. The document can be easily retrieved from a database by specifying merely its title. Documents that are especially suited to SGML standards include form documents, documents where a specific part is frequently modified, documents that are downloaded to different equipment, documents retrieved from a database, and documents written by two or more authors.

7.8.9.2 Computer Graphics Metafile (CGM)

The Computer Graphics Metafile (CGM) is one of the CALS standards for figures, tables, and illustrations. Whereas SGML focuses only on the standard for a document, CGM covers the figures and those documents where figures are included.

7.8.9.3 Initial Graphics Exchange Specification (IGES)

The Initial Graphics Exchange Specification (IGES) specifies the standard for design table produced via CAD. IGES was initially standardized by ANSI and has been implemented as a standard for CALS.

7.8.9.4 Standards for the Exchange of Product (STEP)

The Standards for the Exchange of Product (STEP) model data focus on the manufacturing process. In the manufacturing process, not only design tables but also revised information on the design tables, the configuration of the final product, and sales information have to be taken into consideration. The goal of STEP is to provide the functions by which the companies around the world can easily and freely exchange the design information. Using STEP, the design and manufacturing departments can have concurrent engineering by sharing a database. STEP has been studied under the standardization activities of ISO. Some of them have been standardized. STEP is completely compatible with IGES. In the future, IGES will be included in STEP.

7.9 ELECTRONIC MONEY

A special credit card with an embedded integrated circuit (IC) can store information about deposits and payments. This is what is called *electronic money*. One can deposit funds electronically. For example, salaries can be paid this way. The credit card is, in a sense, a bank passbook. When we buy goods at a department store with electronic money, the amount of the purchase is deducted from the balance of our electronic money account. Salaries are paid into the electronic account. In this way electronic money functions, in effect, like actual money. With the introduction of electronic money, the deposit and payment of money will change drastically from the traditional form.

To establish the electronic money concept requires the development of more reliable equipment for its manipulation. For example, to make it robust and reliable a powerful encryption system has to be established.

The introduction of electronic money will have a significant impact on society. Electronic money will play an important role in this century. The credit card companies, such as VISA International, Inc., and MasterCard International, Inc., are conducting trials with electronic money throughout the world.

In February 2001, NTT Communications, Inc., Japan, announced that it had invented a new type of electronic money, a noncontact type of electronic money in which asymmetric key encryption is installed. A VISA credit card has to be inserted in a machine for a transaction. However, an NTT electronic money transaction does not involve inserting the card in a machine. Rather, you show the card to the machine. NTT Communications was planning to conduct a trial with its card in 2001 in cooperation with partners such as banks, retailers, and railway companies.

In contrast, Sony, Inc., in January 2001, established a new company to handle electronic money, in cooperation with NTT DoCoMo, Toyota Motors, KDDI, Inc., Sakura Bank, and others. Machines are installed at convenience stores, where a customer can transact for noncontact-type electronic money. JR East Japan, Inc., was also planning to conduct a trial with Sony electronic money for its commuters in 2001. Electronic money will surely be put to practical use early in this century.

To accomplish electronic payment, it is very important to be able to transmit information securely over the Internet, information about the purchase order and payment information between a customer and a merchant or between a customer and a credit card company. Secure mechanisms such as Secure Sockets Layer and Secure Electronic Transaction have been developed and installed in the system.

7.9.1 Secure Sockets Layer

First, a customer accesses a server in which a Secure Sockets Layer is installed. Next, the server sends the customer a certificate and a digital signature, issued by a certificating authority, where the digital signature is added to the certificate by using a hash function and the private key of the certificating authority. Then the customer decrypts the digital signature into message digest 1 using the public key of the certificating authority. At the same time, he or she creates message digest 2 from the certificate using a hash function. If message digest 1 and message digest 2 are the same, then the customer's browser can get the public key of the server that is authenticated by the certificating authority. The public key of the certificating authority is installed in the Web server. The Web browser that gets the public key of the Web server generates a session key by using parameters such as time.

The session key is used by the customer and the merchant. The customer encrypts the session key by means of the public key of the Web server and sends it to the server. When the server receives the scrambled session key, it decrypts it by means of its private key. After this interaction between client and server, both the customer and the server share the session key. Using the key, they exchange information securely. For example, the customer encrypts the purchase order by means of the session key and sends it to the server. The server decrypts it by means of the session key and gets the original purchase order. Then the server sends a response to the customer, encrypting it by means of the session key. Then the customer decrypts the scrambled response via the session key and gets the original response.

7.9.2 Secure Electronic Transaction

The method by which a credit card company mediates between a customer and a merchant is important, especially when the customer purchases goods from the merchant over the Internet. It is very important to verify whether the customer and the merchant are the authorized identities. The card company verifies the customer's identity by checking the customer's name, occupation, card ID, and expiration date. Occasionally, the information on a credit card is used illegally, for example, when somebody steals a credit card or card ID, or when somebody opens a business on the network and manages it illegally. These kinds of illegal actions need to be detected and avoided.

In order to manage shopping by credit card securely over the Net, secure electronic transaction (SET) was invented and developed. It specifies the rules about credit card handling among the customer, merchant, and card company. Each of them must get its certificate from a certificating authority before starting the online purchase. While they start an online purchase, the purchase order, such as the name and number of the goods, should be seen exclusively by the merchant. The card information, such as card ID and expiration date, should be seen exclusively by the card company. To achieve this in SET, digital signatures, certificates, and encryption/decryption are introduced. How a customer uses SET is described next, followed by a description of how a merchant and a card company use it.

7.9.2.1 Customer Procedure

First, a customer gets the certificate of the merchant, the certificate of the card company, as well as information about the goods and purchase order by accessing the merchants server over the network. The customer extracts the public key of the merchant and the public key of the company. The certificates verify the merchant's identity and the company's identity. The customer fills out a purchase order and payment information. The purchase order includes the name and number of the goods. Payment information includes the customer's name, street address, age, e-mail address, card ID and expiration date, and the costs of the goods. Next the hash values of the purchase order and payment information are calculated and the values are defined as POh and PIh, respectively. Then POh and PIh are linked and the hash value of the combination calculated and defined as POPIh. Then POPIh is encrypted via the private key of the customer, and the scrambled value is defined as the digital signature POPID. The customer then encrypts PO, PIh, and POPID by means of the public key of the card company. The scrambled values of PO, PIh and POPID are placed in the digital envelope CE. CE is sent to the card company. Then the customer encrypts PO, PIh, POPID, and CE by means of the merchant's public key and gets the scrambled values of PO, PIh, POPID, and CE, respectively. They are put in the digital envelope ME. Then ME is sent to the merchant.

7.9.2.2 Merchant Procedure

When the merchant receives ME from the customer, he or she decrypts it by means of his or her private key and gets PO, PIh, POPID, and CE. First, the merchant calculates the hash value POh of PO and links POh and PIh and calculates the hash value POPIh1 of the combination of POh and PIh. Next, he or she decrypts POPID by means of the customer's public key and gets POPIh2. If POPIh1 and POPIh2 are the same, this confirms that the information transmitted via the network has not been changed illegally and verifies the customer's identity. Then the merchant sends CE to the credit card company and asks the company to verify the customer's identity for an online purchase and to ensure that the credit card is not being used fraudulently by the customer. This procedure enables the merchant to see the customer's purchase order but not the customer's payment information.

7.9.2.3 Credit Card Company Procedure

When the credit card company receives the digital envelope CE, the company decrypts it by means of the company's private key and gets PI, POh, and POPID. It calculates the hash value PIh of PI and the hash value POPIh1 of the combination of POh and PIh. Then it decrypts POPID via the customer's public key and gets POPIh2. If POPIh1 and POPIh2 are the same, this confirms that the information has been transmitted securely and verifies that the customer is an authorized customer. The company checks whether the customer has a bank account and whether he or she can make a purchase by using the customer's PI. The result is sent back to the merchant. When the customer is verified as a customer in good standing with the credit card companies and can pay for the purchase, the company sends the merchant a message that the purchase is allowable. If the customer is found not to be a customer in good standing, the company sends the merchant a message that the purchase is not authorized. This procedure enables the company to see only the payment information but not the purchase order.

7.9.3 Encryption

Encryption is highly important for protecting information from all kinds of attacks while information is being transmitted over the network, such as illegal access by hackers, wiretaps, or dishonest users. Encryption is the only practical way to transmit information securely. Encryption scrambles the information so it cannot be read by anyone who lacks the key. Encryption and decryption are done by means of a key and a key algorithm. By means of the key and the key algorithm, a source scrambles information before transmitting the scrambled data over the network. At the destination, the data is unscrambled with a key to produce the original data.

There are two types of encryption and decryption methods: symmetric key encryption and asymmetric (public) key encryption. Typical encryption and de-cryption are described next.

7.9.3.1 Symmetric Key Encryption Algorithm

With symmetric key encryption, the source text is encrypted by means of a symmetric key and the scrambled text sent to its destination over the Internet. The scrambled text is decrypted into the original text by means of the same key. To accomplish this, the key and the encryption and decryption algorithms have to be shared between the source and its destination. Until recently, this was the widely used method of encryption. With symmetric key encryption, the source and the destination share a common key. The common key is secret, but the algorithm that encrypts and decrypts is not.

- *Data Encryption Standard (DES):* Data Encryption Standard (DES), one of the usual symmetric key encryption methods, is a modification of the best of the symmetric key methods proposed by the National Institute of Standards and Technology (NIST). Its symmetric key is 56 bits long. When text is encrypted, it is broken into components, each 64 bits long. Every component is scrambled by means of the symmetric key, after which all of the scrambled text is sent to its destination over the Internet. At the destination, the scrambled text is decrypted by means of the same key to produce the original text.
- *Improved Data Encryption Algorithm (IDEA):* The improved data encryption algorithm (IDEA) was invented by James Massey and Xuejia Lai of Switzerland in the 1990s. It is said to be faster and more resistant to hacker attack than DES. However, it has not yet been proved better than DES.

• *Fast Encryption Algorithm (FEAL):* The fast encryption algorithm (FEAL) was invented by NTT, Japan, in 1985. Its key length is 64 bits. The text is broken into 64-bit-long components and each component encrypted and decrypted the same way at source and destination, respectively. The several versions of FEAL (FEAL-4, FEAL-8, FEAL-16, and FEAL-32) differ in the number of times the algorithm is applied. In the case of FEAL-4, for example, the encryption algorithm is applied four times to get the scrambled text.

7.9.3.2 Problems with Symmetric Key Encryption

The major problem with symmetric key encryption is how to keep the symmetric key secret and how to send it to the destination safely, without attack. There is a great possibility that the key could be stolen by a hacker while sending it to the destination over the Internet. To avoid this, unencrypted text should never be transmitted without first being scrambled. Another problem is that the number of keys is increasing as the number of customers increases.

7.9.3.3 Public Key Encryption Algorithm

To overcome the problems with the symmetric key method, the asymmetric (public) key encryption algorithm was invented by W. Diffie and M. E. Hellman of Stanford University in the 1970s. In this method, different keys are used for encryption and decryption. The encryption key is public, but the decryption key is private and secret. The decryption key is a prime factor obtained from factoring the encryption key. That is, the number of keys is twice the number of customers. The customer keeps his or her own key secret.

With the Rivest–Shamir–Adleman encryption method (RSA), which was invented by R. L. Rivest, A. Shamir, and L. Adleman in 1977 and based on the Diffie–Hellman method, it takes a lot of time to find a prime factor from the public key. For example, for a number with 200 digits it would take over 380,000 years to discover a prime factor, even using a supercomputer. Thus, finding a prime factor is practically impossible. This is the principle behind the RSA encryption method. Its major defect is that encryption and decryption take more time than with the symmetric key method.

7.9.3.4 Hybrid Method

The hybrid method is a combination of symmetric and asymmetric key encryption. In this method, a public key is given. A symmetric key is encrypted by means of the public key. At the same time, the text is encrypted with the symmetric key. Then the scrambled symmetric key and the scrambled text are sent to the destination over the Internet. At the destination, using the private key that corresponds to the public key, the scrambled symmetric key is decrypted to produce the original symmetric key. Then the scrambled text is decrypted using the symmetric key. Finally, the original text is obtained.

8

INTELLIGENT COMMUNICATION SYSTEMS

As the importance of the communication network in our society increases, the more human-friendly telecommunication services must become. As the communication network becomes increasingly indispensable, it becomes increasingly necessary to make the network human-friendly by applying AI, such as knowledge processing and natural language processing, to our telecommunication services. To accomplish this, research and development in the following areas is being carried out:

- Telephones that make a phone call based on name and/or affiliation rather than a phone number
- Interpreting telephones that perform real-time language interpretation
- Directory service that retrieves a phone number based on the subscriber's name or job category
- Virtual space teleconferencing system that produces a human-oriented communication environment with realistic sensations

This chapter describes an intelligent communication system that provides such human-friendly services.

8.1 CONCEPT OF INTELLIGENT COMMUNICATION SYSTEMS

An intelligent communication system provides human-friendly telecommunication services by applying intelligent processing. Examples of intelligent telecommunication services include: intelligent switching services, such as the intelligent phone or one-touch dialing; intelligent directory service, such as the retrieval of a phone number based on the subscriber's name or job category; and intelligent communication processing services, such as the interpreting telephone or media conversion. The intelligent communication system is structured hierarchically, that is, in a layered structure. The functions of each layer are as follows:

- The *physical layer* includes the electrical and physical conditions of the network.
- The *data link layer* includes the rules for communication between neighboring nodes.
- The *network layer* describes the rules for communication between the source node and the destination node.
- The *transport layer* includes the rules for communication between the source process and the destination process.
- The *session layer* includes how to establish the communication link between the source process and the destination process and how to communicate (e.g., duplex communication or half-duplex communication).
- The *presentation layer* includes code conversion.
- The *application layer* includes the application functions provided to the user.

In the intelligent communication system the intelligent processing layer is defined as above the application layer. The user program is defined as above the intelligent processing layer. The hierarchical structure of Intelligent Communication System is shown in Figure 8.1.

8.2 FUNCTIONS OF THE INTELLIGENT PROCESSING LAYER

The functions of the intelligent processing layer are performed by the following systems:

- Knowledge-base system
- Natural language processing system
- Media conversion system

The knowledge-base system provides problem-solving facilities by using knowledge in a knowledge base. The natural language processing system provides interpreting facilities and language understanding. The media conversion system

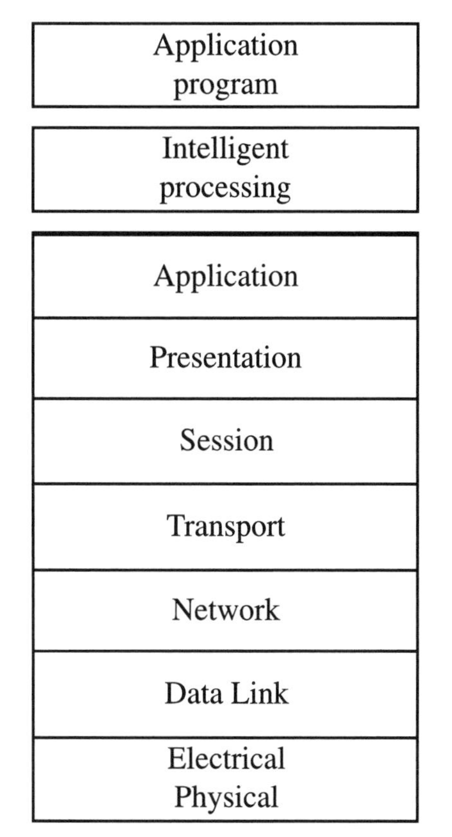

FIGURE 8.1 Hierarchical structure of intelligent communication system.

provides speech-to-character conversion, sentence-to-image conversion, and vice versa.

8.3 STRUCTURE OF THE KNOWLEDGE-BASE SYSTEM

The knowledge-base system provides problem-solving facilities by using domain knowledge stored in a knowledge base. As shown in Figure 8.2, the knowledge-base system comprises a knowledge base, an inference engine, an inference process explanation module, a knowledge-base editor, and a human–machine interface module.

• The *knowledge base* stores domain knowledge for problem solving. For example, in order to construct a knowledge-base system for medical diagnosis, information about a disease and its symptoms is input. This is domain knowledge for medical diagnosis.

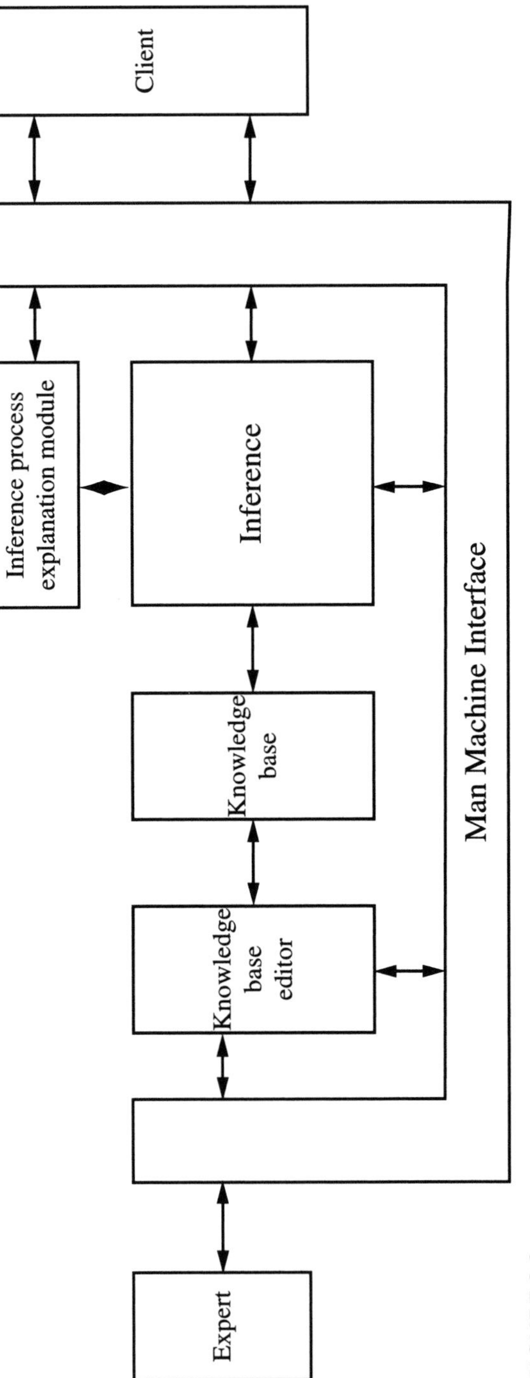

FIGURE 8.2 Knowledge-base system.

• The *inference engine* solves a problem by using domain knowledge. For example, using the domain knowledge for medical diagnosis, a problem is solved about what disease is expected for a given symptom.

• The *inference explanation module* explains the problem solving process, that is, why a particular disease is expected from the given symptom.

• The *knowledge-base editor* provides the facilities for inputting information and editing or maintaining the knowledge base. For example, a set of a symptom and its disease is input by using the knowledge-base editor.

• The *human–machine interface module* interprets users' requests to the system, given in written language, images, spoken language, or gestures. The module interprets the requests and translates them into machine-understandable form.

9

DESIGN METHODOLOGY FOR TELECOMMUNICATION SERVICES

As already mentioned, information technology will play an important role in the coming century. As IT advances in our society, a variety of telecommunication services will be desired. In response to these demands, easy-to-use and effective telecommunication services will become available. The newly developed services will be provided not only by common carriers but also by value-added companies. To realize such services it is desirable to establish human-friendly development environments. In this chapter the design methodology for telecommunication services is described.

9.1 STATE-OF-THE-ART DESIGN METHODOLOGY

The design and development of telecommunication services are carried out as follows:

(1) Design of the telecommunication service
(2) Description of the telecommunication service by signal sequence or by functional Specification and Description Language (SDL)

(3) Design of the program structure
(4) Programming in C language and debugging

In process (1), the telecommunication service is designed. In process (2), description by signal sequence or SDL is performed, depending on the design sheet. In process (3), the program structure is designed according to the description. In process (4), programming is done and the programs debugged. An example of the signal sequence is shown in Figure 9.1. In the figure an offhook signal is issued from the user to the system. The system asks the user which number is to be called, the user responds by giving a number, in this case 03-5286-3830, and the system reconfirms the number to the user. The calling process is performed in this manner. The example is described in SDL in Figure 9.2. The initial state is the idle state. The user picks up the phone and the system asks which number to call. The system waits for a response from the user. The user inputs the

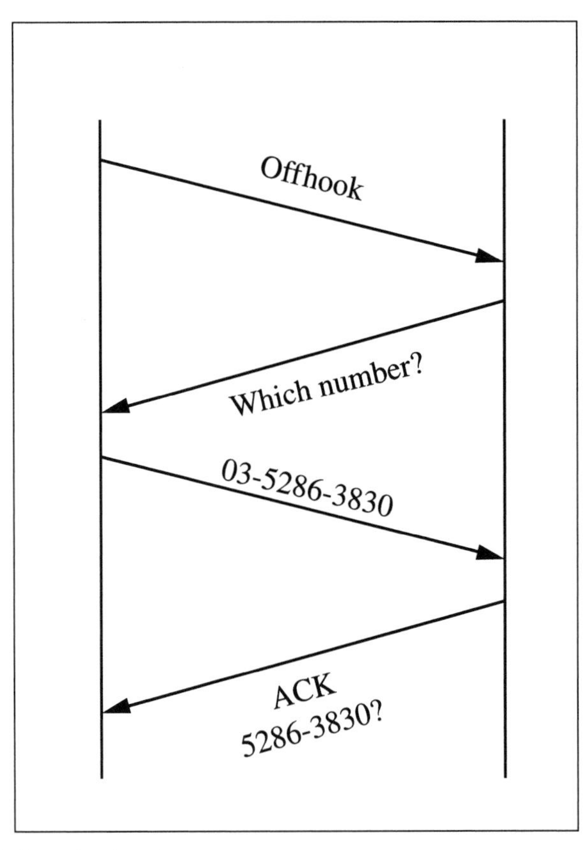

FIGURE 9.1 Example of signal sequence.

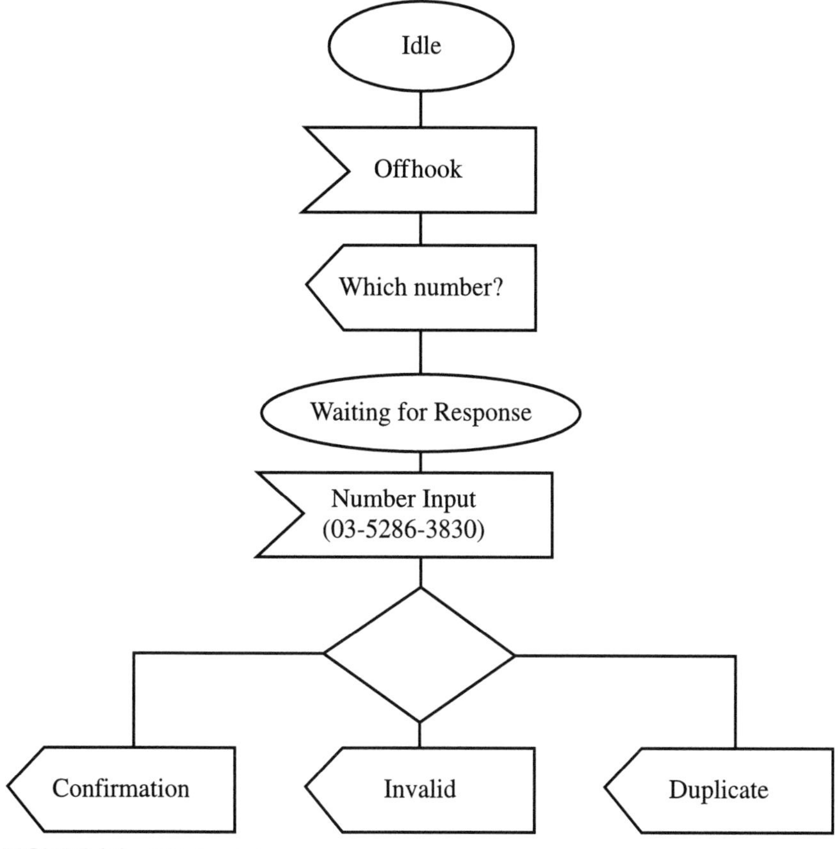

FIGURE 9.2 SDL description.

number for calling, 03-5286-3830. Depending on the input by the user, the system chooses "confirmation," "invalid," or "duplicate" and responds accordingly to the user.

The signal sequence description or SDL description corresponds to the calling process. Only professionals can describe the process in terms of signal sequence or SDL. To further promote IT in our society, the design method needs to be easy to manipulate not only by professionals but also by nonprofessionals. Development of an easy-to-use design method will facilitate the smooth adoption of IT by society. One of the advanced design methods is described next.

In this method the design process is performed as follows:

(1) Design of telecommunication service
(2) Description by state transition rule (STR)
(3) Simulation

This is different from the method that uses signal sequence or SDL. The characteristics of this method are:

(1) The STR description is introduced.
(2) The calling process is simulated by the STR simulator.

Once a service is described in STR, the process is simulated immediately. It is not necessary to describe all of the service procedures to simulate the calling process. In this way rapid prototyping is performed with ease.

9.2 DEFINITIONS

Telecommunication services are described in terms of state and operation. For example, plain old telephone service (POTS) is described as follows.

(1) *Description of state:* The idle state of the telephone is described as

idle(A)

This means that telephone A is in an idle state. The state in which a dialtone is heard at telephone A is described as

dialtone(A)

In this way, a state is described by using the predicate.

(2) *Description of operation:* When telephone A calls telephone B, this is defined as

dial(A, B)

To describe that telephone A is answered, we write

offhook(A)

(3) *Description of telephone service:* Here are some examples of telephone service descriptions.

Example1: idle(A) offhook(A) : dialtone(A)

This means that when telephone A is idle and it is taken offhook, then a dialtone is heard at telephone A.

Example 2: dialtone(A), idle(B) dial(A, B) : ringback(A, B), ringing(B, A)

This means that a dialtone is heard at telephone A and telephone B is idle. When telephone A calls telephone B, telephone B rings and a ringback tone is heard at telephone A. This process is shown schematically in Figure 9.3.

In general, a telephone onhook is the initial state. When the phone is taken offhook or calls another phone, the phone gives a dialtone or a ringback tone. Thus, when an event occurs at the phone, the state is transferred to the next state. The transition of a state is shown in Figure 9.4.

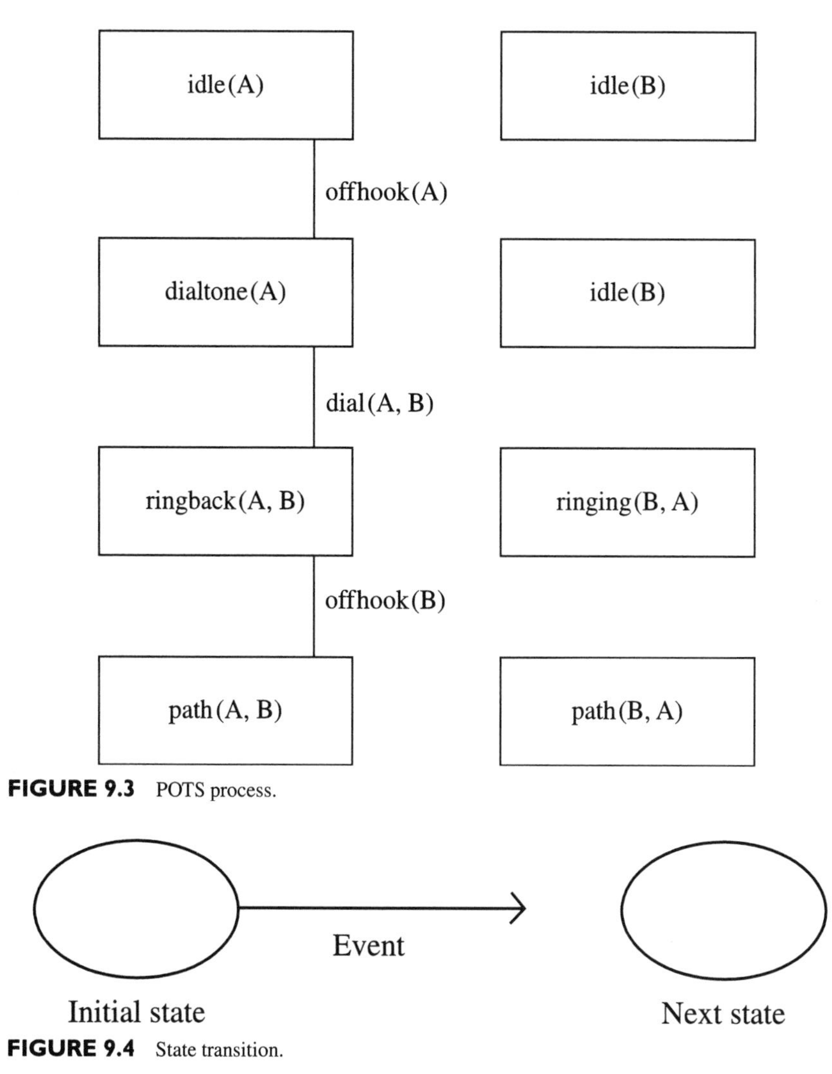

FIGURE 9.3 POTS process.

FIGURE 9.4 State transition.

According to graph theory, this state transition is a *directed graph*, that is, a graph where nodes are linked by a directed branch.

9.3 GRAPH THEORY

(1) A graph is composed of a node and a branch.
(2) A graph that has a directed branch is a directed graph (see Figure 9.5). In the directed graph where node i is linked to node j by a directed

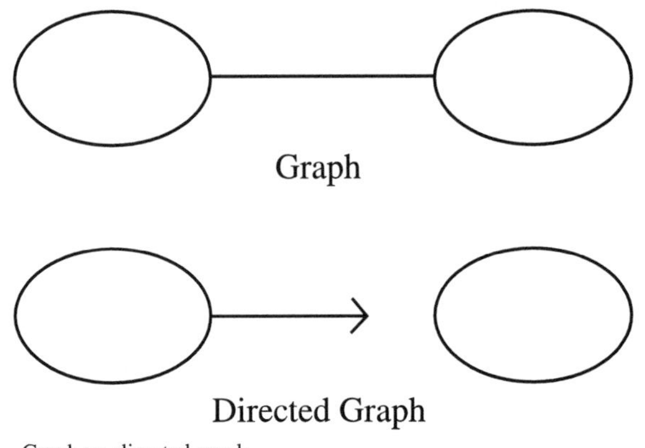

FIGURE 9.5 Graph vs. directed graph.

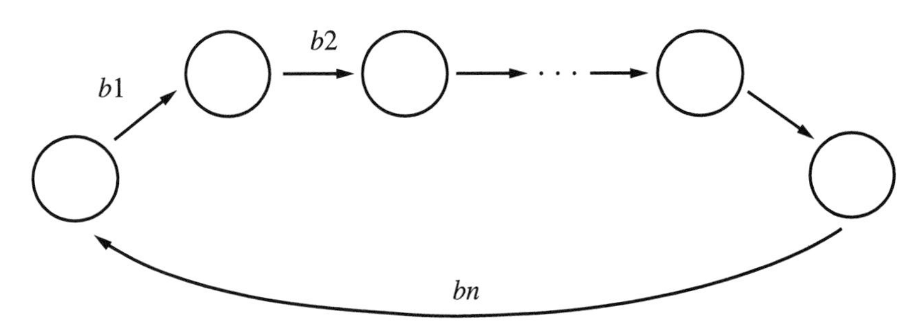

FIGURE 9.6 Cycle example.

branch, node i is the *parent* node of node j and node j is the *child* node of node i.

(3) Regardless of the direction of a branch, the sequence of the branches $b1, b2, \ldots, bn$ that are connected together is called the *walk* of length n and is defined as $(b1, b2, \ldots, bn)$.

When there is a sequence of nodes $(b1, b2, \ldots, bn)$ linked by directed branches and the ending point of bi is the beginning point of $b(i+1)$, then the sequence is called a *path* of length n. When there is a path $(b1, \text{b}2, \ldots, bn)$ and $b1$ and bn are the same, the path is called a *cycle* (Figure 9.6). If there is a walk for every node in a graph, the graph is *connected*. The connected graph that has no cycle is called a *tree* (Figure 9.7). The tree that has only one root is called a *principal* tree. If there are two paths $p1$ and $p2$, where the first node of path $p1$ and that of path $p2$ are the same, the last node of path $p1$ and that of path $p2$ are the same, and any other nodes except the first node and the last node are different, then the paths are called a *circuit* (Figure 9.8). A graph is shown in Figure 9.9.

The procedure for detecting the goal node by starting from the initial node is called *problem solving* by using a state space. Let's study an example of problem solving.

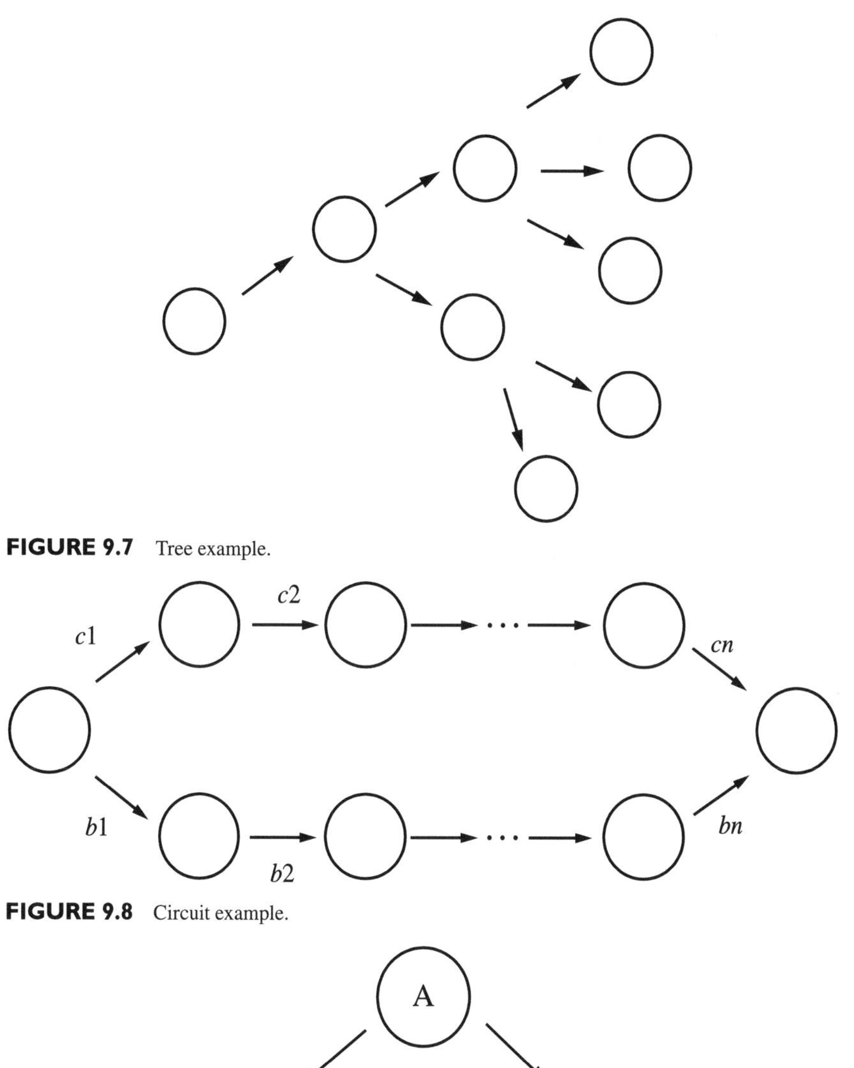

FIGURE 9.7 Tree example.

FIGURE 9.8 Circuit example.

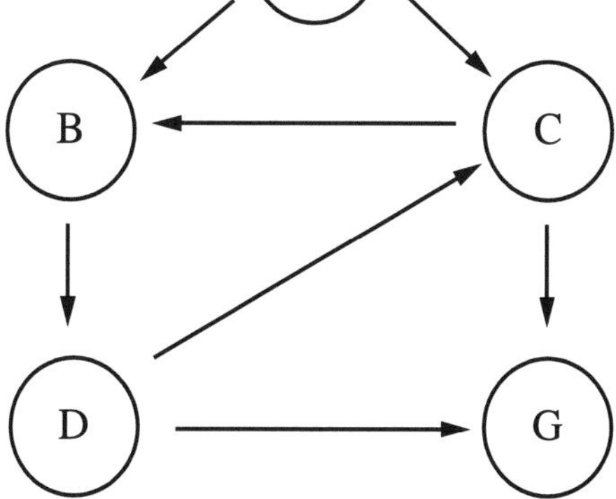

FIGURE 9.9 Graph example.

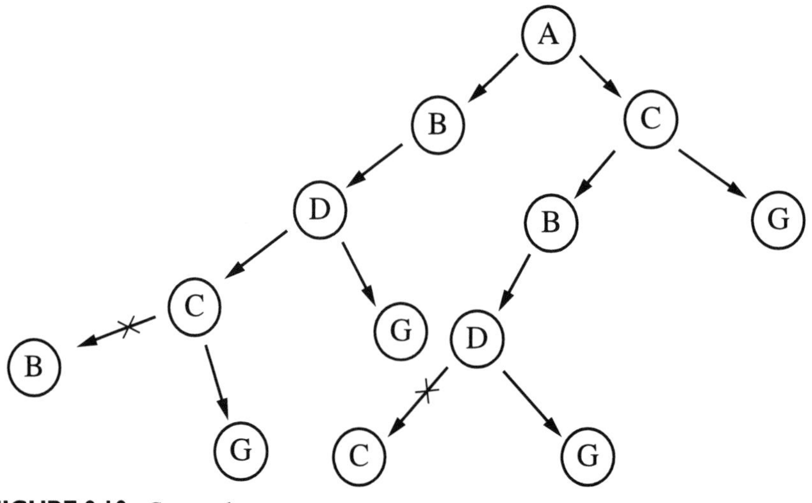

FIGURE 9.10 Generated tree.

The problem is to determine a path connecting node A and node G. This is solved as follows.

(1) Any nodes linked by directed paths from node A are detectable; node B and node C are detected.
(2) In the same way, any nodes linked by directed paths from node B and node C are detected.
(3) In this way, the nodes linked one to another without a cycle are detected.
(4) Then the tree structure shown in Figure 9.10 is obtained.
(5) All of the paths from node A to node G form the solution as follows.

$A \rightarrow B \rightarrow D \rightarrow C \rightarrow G$

$A \rightarrow B \rightarrow D \rightarrow G$

$A \rightarrow C \rightarrow B \rightarrow D \rightarrow G$

$A \rightarrow C \rightarrow G$

9.4 EXAMPLE DESCRIPTION OF TELECOMMUNICATION SERVICES

An example of a description of POTS is shown in Figure 9.11. The description of POTS in the state space is shown in Figure 9.12.

The meaning of the predicates in Figures 9.9 and 9.10 is as follows.

pots-1)	idle(A) offhook(A) : dialtone(A).
pots-2)	dialtone(A), idle(B) dial(A, B) : ringback(A, B), ringing(B, A).
pots-3)	ringback(A, B), ringing(B, A) offhook(B) : path(A, B), path(B, A).
pots-4)	path(A, B), path(B, A) onhook(A) : idle(A), busy(B).
pots-5)	dialtone(A), not [idle(B)] dial(A, B) : busy-dial(A, B).
pots-6)	dialtone(A), onhook(A) : idle(A).
pots-7)	ringback(A, B), ringing(B, A) onhook(A) : idle(A), idle(B).
pots-8)	busy(A) onhook(A) : idle(A).
pots-9)	busy-dial(A, B) onhook(A) : idle(A).
pots-10)	dialtone(A) timer(dialtone(A)) : howler(A).
pots-11)	busy-dial(A, B) timer(busy-dail(A, B)) : howler(A).
pots-12)	busy(A) timer(busy(A)) : howler(A).
pots-13)	howler(A) onhok(A) : idle(A).
pots-14)	timer : dialtone(A) 5sec.
pots-15)	timer : busy-dial(A, B) 5sec.
pots-16)	timer : busy(A) 5sec.

FIGURE 9.11 POTS description.

path(A, B) means the establishment of a path between A and B. Then a caller and the called party can communicate by talking or exchanging information.
path(B, A) means the establishment of a path between B and A.
idle(A) means that terminal A is idle.
busy(B) means that terminal B is busy.
not [idle(B)] means that terminal B is not idle.
busy-dial(A, B) means that terminal A hears a busy tone from terminal B.
timer [dialtone(A)] is an event, and it means that a dialtone is heard at terminal A and a timer works at terminal A.
howler(A) means that a howler tone is heard at terminal A.
timer is defined in the lines for pots-14, -15, -16 in Figure 9.11. For example, at line pots-14, a timer works in 5 seconds in the dialtone(A) state.

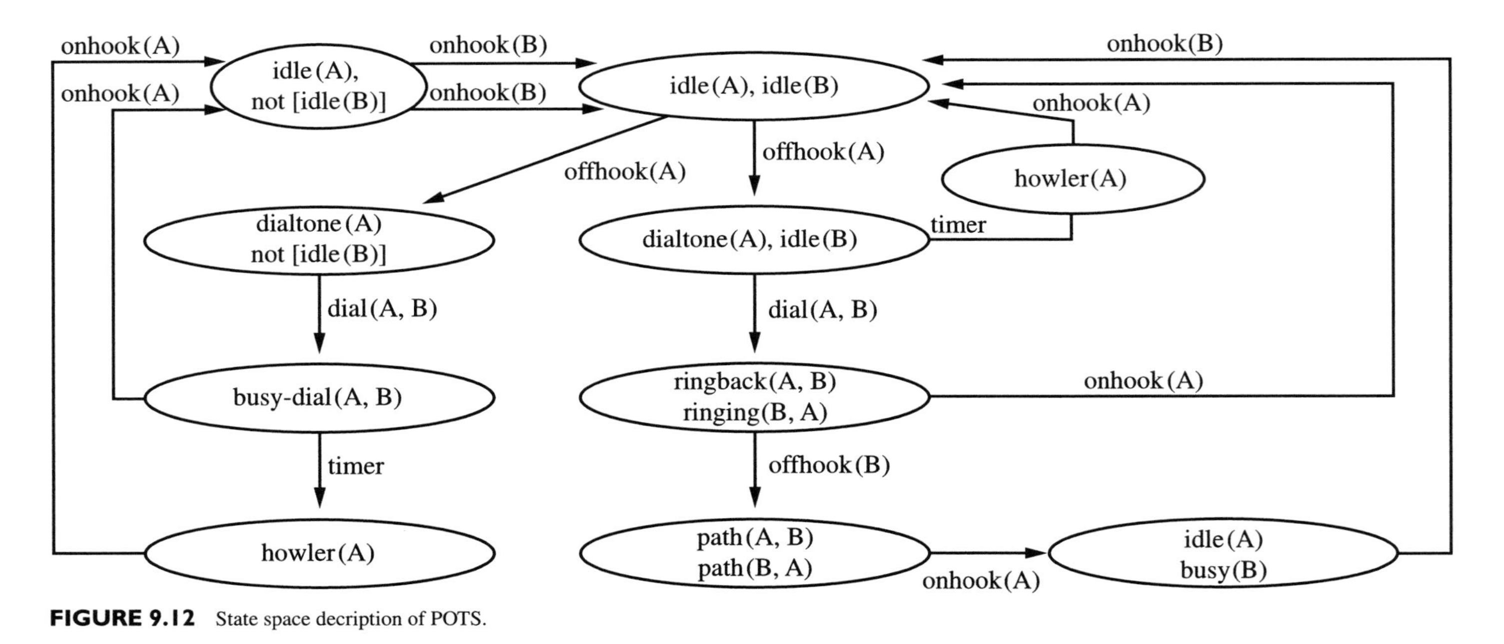

FIGURE 9.12 State space decription of POTS.

9.5 CONFLICTS AMONG TELECOMMUNICATION SERVICES

When telecommunication services are described in predicate logic, there are possibilities for conflict with other telecommunication services. In this section, examples of conflict and conflict resolution are discussed.

(1) *Call waiting service* (Figure 9.13): When terminal A calls terminal B while terminal B and terminal C are talking to each other, the system lets terminal B know of the arrival of the call. This is call waiting service.

(2) *Call forwarding variable service* (Figure 9.14): When terminal A calls terminal B while terminal B is busy, the call is forwarded to terminal D, which has been designated by terminal B as the forwarded terminal.

(3) *Conflict resolution* (see Figure 9.15 for examples of service descriptions that come into conflict and the rules by which such conflicts are resolved): When an event dial(A, B) occurs, system checks states A and B. State A is a dialtone for both (1) and (2) and state B is idle(B) for both (1) and (2). Initial states (1) and (2) are the same for the event dial(A, B), so the system cannot choose one rule in a consistent manner. This requires conflict resolution.

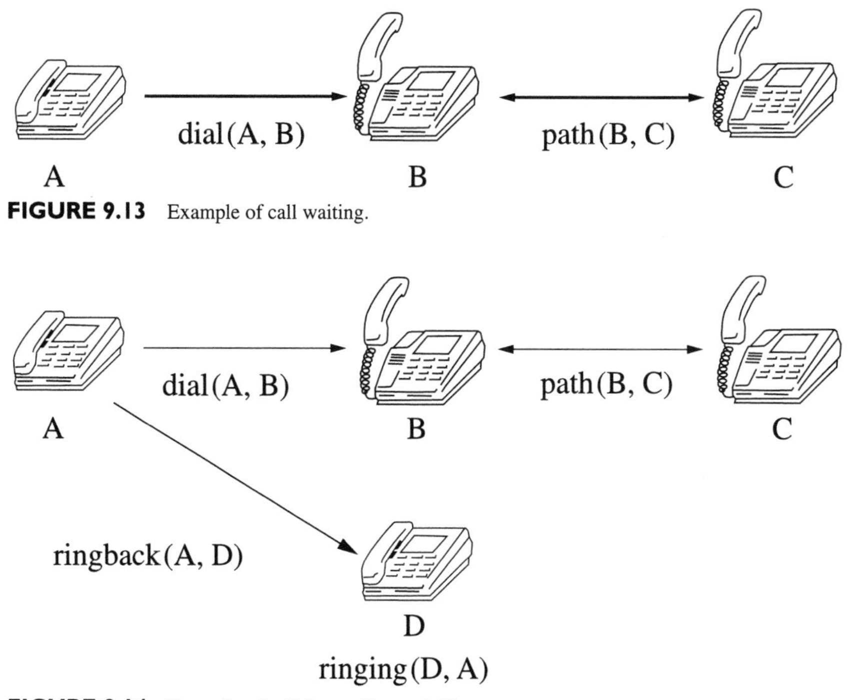

FIGURE 9.13 Example of call waiting.

FIGURE 9.14 Example of call forwarding variable.

(CW : Call Waiting)
dailtone(A), m-cw(B), path(B, C)
dial(A, B) :
ringback(A, B), cw-ringing(B, A), m-cw(B), path(B,C).
(CFV : Call Forwarding Variable) :
dialtone(A), path(B, C), m-cfv(B, D), idle(D)
dial(A, B) :
ringback(A, D), ringing(D, A), path(B, C), m-cfv(B, D).

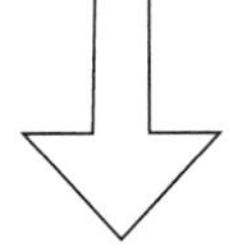

A New Rule where CFV is applied
dialtone(A), path(B, C), m-cfv(B, D), m-cw(B), idle(D)
dial(A, B) :
ringback(A, D), ringing(D, A), path(B, C), m-cfv(B, D), m-cw(B).

FIGURE 9.15 Conflict resolution.

In Figure 9.15, the following predicates are used.

m-cw(B) means that a call waiting service is defined at terminal B.
cw-ringing(B, A) means that terminal A is calling terminal B and a ringing tone is heard at terminal B because of call waiting service.
m-cfv(B, D) means that a call is transferred to terminal D when terminal B is busy.

In this case a call forwarding variable is chosen. In general, a conflict occurs when the initial states and operations of two or more rules are the same. In examples (1) and (2), the initial states of (1) and (2) are dialtone(A), path(B, C) and the operations of (1) and (2) are dial(A, B). The rules that do not come into conflict are:

Rule (1): dialtone(A), idle(B) dial(A, B) : ringback(A, B) ringing(B, A)
Rule (2): dialtone(A), not [idle(B)] dial(A, B) : busy-dial(A, B)

The initial states of the foregoing descriptions are not the same, which means there is no conflict. In general, rules that come into conflict are as follows:

For the event ev(A, B), the initial states A and B are the same and the next states are defined as follows.
For terminal A, the next states are {PA1(A, B), PA2(A, B),…, PAm(A, B)}.
For terminal B, the next states are {PB1(B, A), PB2(B, A),…, 1PBn(B, A)}.

This means that states PA1(A, B), PA2(A, B), …, PAm(A, B) may occur simultaneously at terminal A. In the same way, for terminal B, states PB1(B, A), PB2(B, A),…, PBn(B, A) may occur simultaneously. Therefore conflict resolution is needed.

9.6 CONFLICT OF CHARGE POLICY

When more than one service applies to a terminal, a conflict of charge policy may occur. Examples of such conflicts follow.

Case 1: toll-free phone service. In Figure 9.16 terminal B is charged when a toll-free phone call is issued to terminal B.

Case 2: Call forwarding variable service. In Figure 9.17, call forwarding service is defined at terminal B and B is busy. When terminal A calls terminal B, the call is forwarded to terminal C. Then the call between terminal A and B is charged to terminal A, and the forwarded call between terminal B and terminal C is charged to terminal B. When toll-free phone service and call forwarding variable service are defined at terminal B, then whether a call between A and B is charged to A or B is not clearly decided. In this case the charge policy has to be rewritten to resolve the conflict.

In general the specification for the charge policy is described as follows:

IF C THEN T, R

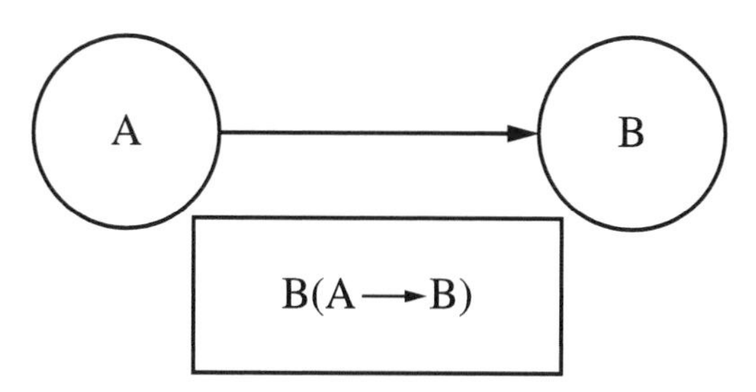

FIGURE 9.16 Toll-free phone charge.

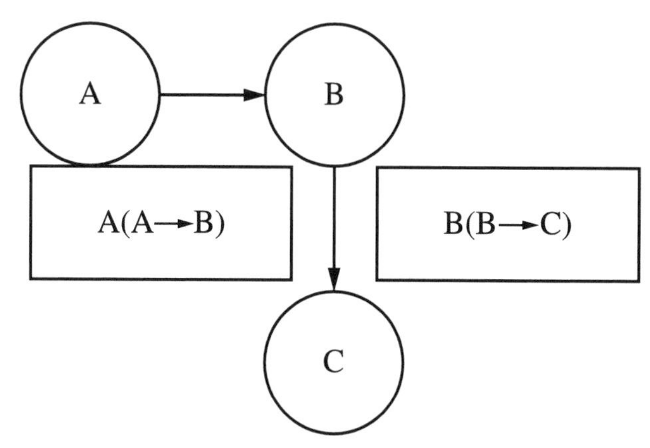

FIGURE 9.17 Call forwarding charge.

where C is conditional and T and R are actions. C comprises a state description and an event. T comprises the terminal to be charged and the domain to be charged. R is a fee. Here are some examples.

Example 1: A call forwarding variable is defined at terminal B. As mentioned earlier, a call between terminal A and terminal B is charged to terminal A and a forwarded call between terminal B and terminal C is charged to terminal B, where the fee is r1. The charge policy is then described as follows:

IF m-cfv(B, C), dial(A, B) THEN A(A-B), B(B-C), r1,

where m-cfv(B, C) means that a call to terminal B is transferred to terminal C when terminal B is busy.

Example 2: For conventional phone calls, when terminal A calls terminal B, then the call is charged to terminal A. The charge policy in this case is described as follows:

IF dial(A, B) THEN A(A-B), r2.

In general, rules that come into conflict are shown here. The following rules are given:

Rule (1) IF S1(A, B) E1(A, B) THEN B(A-B), r1.

Rule (2) IF S2(A, B) E1(A, B) THEN A(A-B), B(B-C), r2.

For event E1(A, B), there are two rules, and the system cannot select one of them. Therefore these two rules have to be rewritten to resolve the conflict.

9.7 HIGH-LEVEL DESCRIPTION OF TELECOMMUNICATION SERVICES

(1) A basic function of telecommunication services is to connect or disconnect a line for a terminal.

(2) Telecommunication services are described in state transition rules as follows:

Condition (state1, action) and result(state 2)

When a terminal is in state1 and action is taken, state1 is transferred to state 2.

(3) The state of the terminal is defined as follows.
State of service subscription: subscribed/not subscribed
State of service activation: active or inactive
Line state: connecting, holding, or disconnecting
Terminal state: active or inactive
Response from service: response by user number or command input
Relation among terminals: person-to-person call, call forwarding variable, or 3-party call

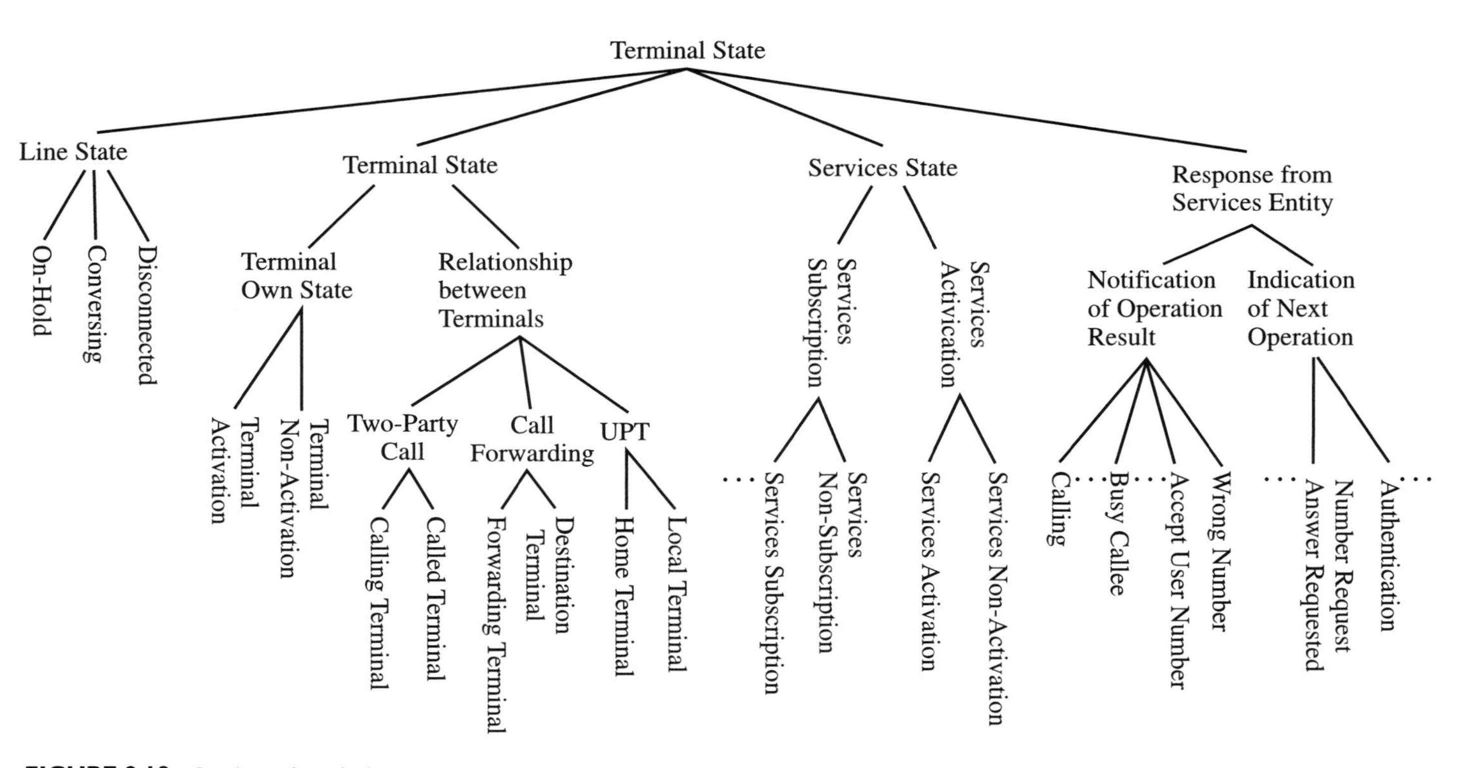

FIGURE 9.18 Ontology of terminal states.

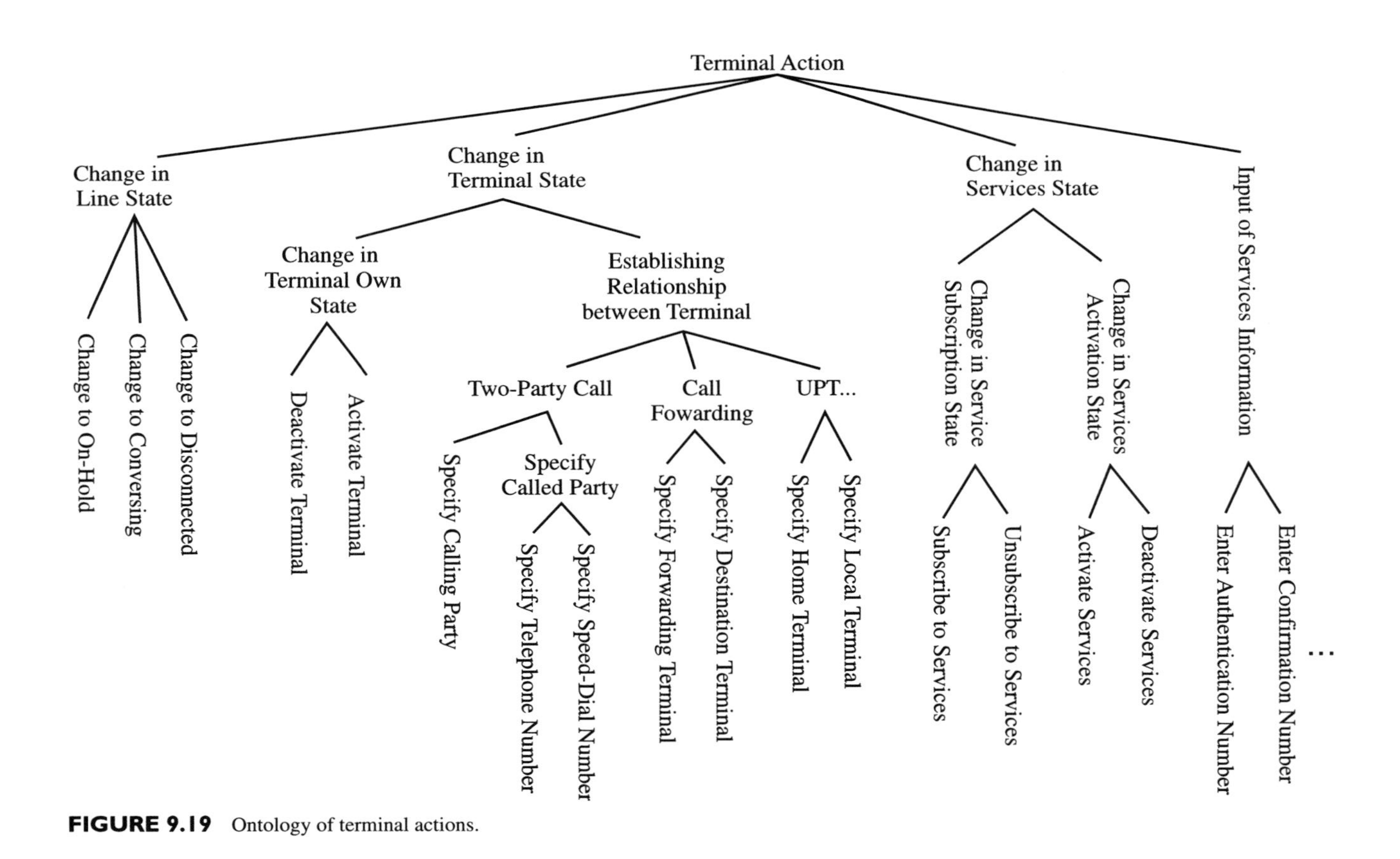

FIGURE 9.19 Ontology of terminal actions.

(4) Action is defined as follows.

Input of service information: input of user number, input of authentication

Change of service state: activation or deactivation of service subscription or nonsubscription of service

Change of terminal state: universal personal telephone (UPT), local terminal, and home terminal

- Call forwarding variable: forwarding terminal and forwarded terminal
- Person-to-person call: caller terminal and called terminal
- Change of terminal: terminal activation or deactivation
- Change of line state: converse, disconnect, or hold

All of these terminal states are listed and stored as an ontology of terminal states.

9.8 REQUIREMENT SPECIFICATION

(1) The requirement specification is described in natural language as follows. When the initial states are Si{, Si} and an action *a* is taken, the states are moved to So{, So}, where Si and So are states, *a* is an action, and {⧅} shows iteration.

(2) The requirement specification form is interpreted as in the following example: When onhook(A) is performed at the terminal state path(A, B), then the terminal state is moved to idle(A). The validity of Si and So is checked by using the ontology of the terminal states shown in Figure 9.18. The validity of *a* is checked by using the ontology of the terminal actions shown in Figure 9.19. If Si, So, and *a* are valid, then the form is translated to an STR form.

10

BASIC TECHNOLOGY OF THE INTELLIGENT COMMUNICATION SYSTEM

Intelligent telecommunications occur as follows.

(1) When a breakdown of or some congestion in a communication network occurs, an action is taken to recover from the breakdown or congestion. This requires knowledge of the breakdown or congestion, called a *production rule.*

(2) For the design of telephone services, it is necessary to know about terminal states or line states. *Terminal states* include the person-to-person, three-party call, call forwarding variable, and call waiting states. *Line states* include the conversing, on-hold, and disconnection states. When a terminal is called and that terminal has call forwarding variable service, the call is forwarded to the designated terminal. The states are defined in a semantic network.

(3) Intelligent communication services include intelligent directory service, intelligent switching service, and intelligent inquiry service. With intelligent directory service, the system, when asked in natural language for the phone number of, say, Mr. Tanaka who lives in San Francisco, supplies the requested number. With intelligent switching, service the system, when asked, say, "Please call Mr.Tanaka,"

connects the call to Mr. Tanaka. With intelligent inquiry service, the system, when asked, say, "Please let me know of a cut-rate clothing shop in San Francisco," supplies the requested phone number.

With all these services, the system interprets the inquiry from a client and returns an answer. There are a variety of requirements from clients, expressed in a variety of utterances. It is difficult for a computer to understand every such utterance. Predicate logic is one way to handle these problems, and predicate logic is appropriate not only for human beings but also for computers. The representations of knowledge, such as production rules and semantic networks, and the inference machine, also useful for telecommunications, are described in this chapter, as are propositional logic and predicate logic.

10.1 APPLICATION OF PRODUCTION RULES TO TELECOMMUNICATIONS

Production rules are used to describe the maintenance and operation of a communication system. A *production rule* (PR) is defined as follows:

IF *conditional clause,* THEN *action clause.*

An execution algorithm for a PR is shown in Figure 10.1. Here's a strategy for executing a production rule:

(1) A production rule that is matched to a variable in working memory (WM) is chosen.
(2) A production rule with the longest conditional clause is chosen.
(3) All of the production rules that are matched to the variables in WM are chosen.
(4) All of the production rules that are matched to the variables in WM except the production rules that have been chosen so far are selected.

Example 1—Execution of Production Rules:

(1) In Figure 10.2, "Pork" and "Cabbage" are in WM.
(2) Production rules P1 and P5, which are matched to "Pork" and "Cabbage," are chosen.
(3) P5 is selected first and executed. After execution, "Vegetable" is derived and stored in WM.
(4) P6, which is matched to "Vegetable," is chosen.
(5) P6 is executed. After execution, "Ted dislikes" is stored in WM. No production rule matched to "Ted dislikes" is found, so execution is complete and the conclusion is that "Ted dislikes cabbage."
(6) P1 is executed and "Meat" is stored in WM.

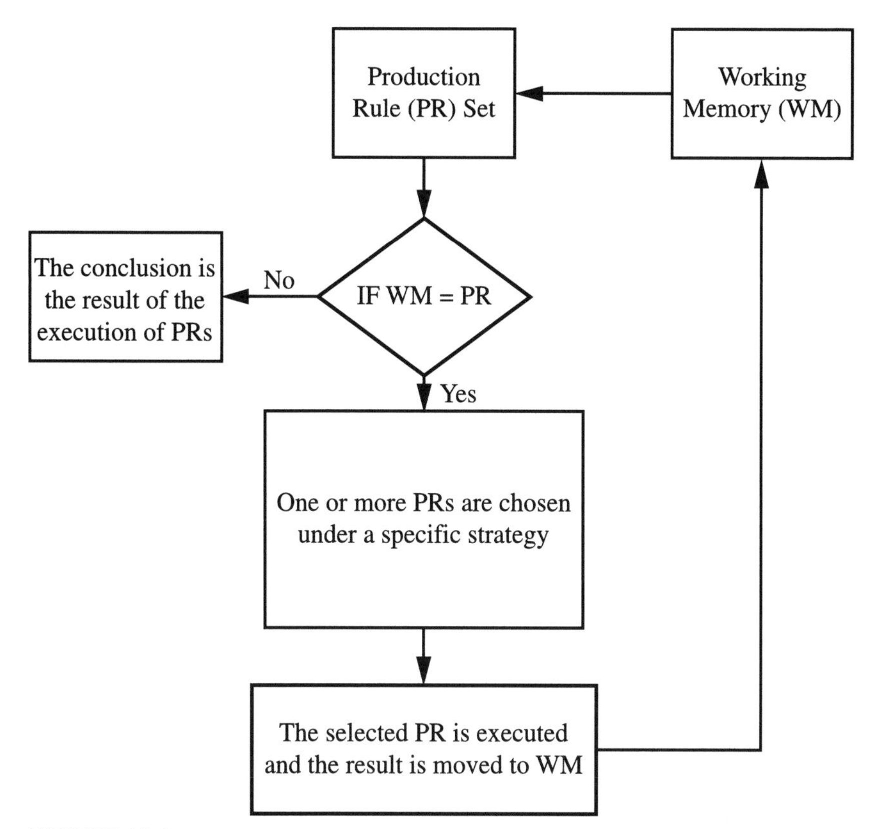

FIGURE 10.1 PR inference.

(7) Production rule P2, which is matched to "Meat," is chosen and executed. After execution, "Ted likes" is found and execution is complete. The conclusion is that "Ted likes meat." The process flow is shown in Figure 10.2.

Example 2—Finding an Alternative Rule: The following rule is given:

IF A is busy, THEN the traffic goes through B.
IF both A and B are busy, THEN the traffic goes through C.

A routing example is shown in Figure 10.3.

Knowledge about a store can be presented in a semantic network, as in Figure 10.4. The same knowledge can be described with production rules, as follows:

(1) IF a shop is perishables shop A, THEN it is a perishable shop.
(2) IF a shop is a perishables shop, THEN it is a food shop.
(3) IF a shop is a food shop, THEN it is a shop.
(4) IF a shop is a candy shop, THEN it is a food shop.

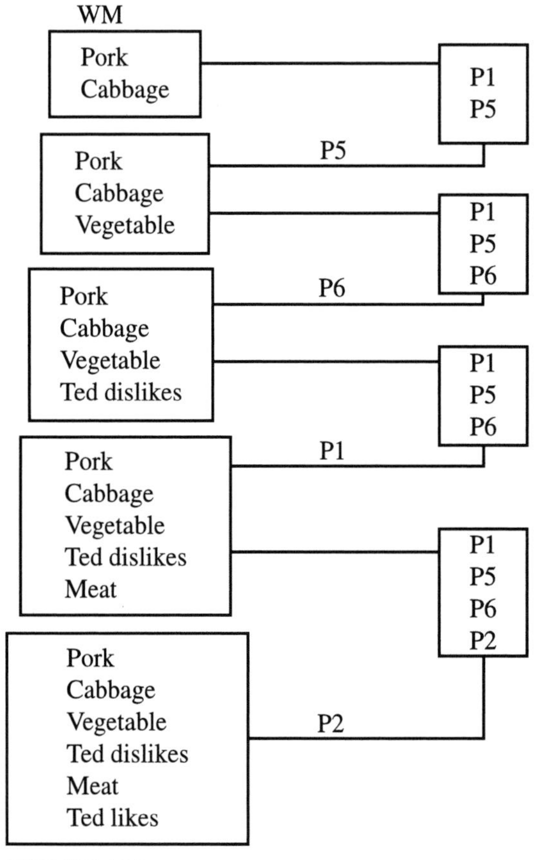

FIGURE 10.2 PR process flow.

(5) IF a shop is a grocery shop, THEN it is a food shop.
(6) IF a shop is a tailor's shop, THEN it is a clothing shop.
(7) IF a shop is a clothing shop, THEN it is a shop.

Problem: Is perishables shop A a shop?

(1) Perishables shop A is stored in WM.
(2) PR (1) is chosen, which is matched to "Perishables shop A."

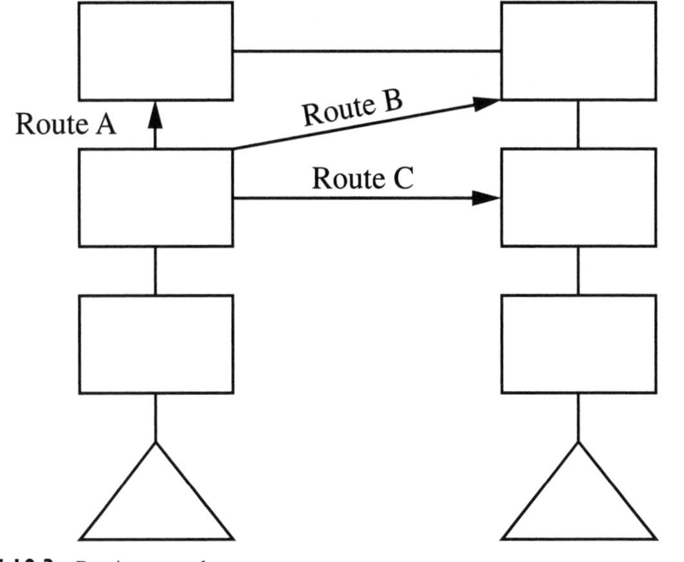

FIGURE 10.3 Routing example.

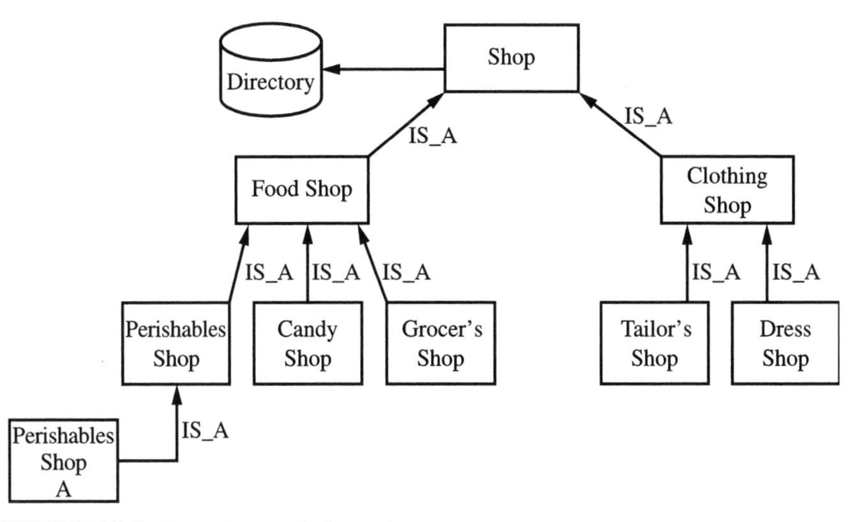

FIGURE 10.4 Semantic network about a shop.

(3) After execution of PR (1), "Perishables shop" is derived and stored in WM.
(4) Then PR (2), which is matched to "Perishables shop" is chosen and executed. After execution, "Food shop" is derived and stored in WM.
(5) Next PR (3), which is matched to "Food shop," is chosen and "Shop" is derived.

Since "Shop" is derived from "Perishables shop A," the answer to our problem is yes.

10.2 DESCRIPTION OF TELECOMMUNICATION SERVICES IN A SEMANTIC NETWORK

Semantic networks can be used to describe a terminal state and a line state. A *terminal state* is shown in Figure 9.18. Here's one interpretation of this semantic network:

Person-to-person call PART_OF relationship between terminals.
Three-party call PART_OF relationship between terminals.
Call forwarding variable PART_OF relationship between terminals.

In other words, there is a person-to-person call, a three-party call, and a call forwarding variable in a relationship between terminals.

Here's another interpretation of this semantic network:

Person-to-person call IS_A relationship between terminals.
Three-party call IS_A relationship between terminals.
Call forwarding variable IS_A relationship between terminals.

In other words, a person-to-person call is a relationship between terminals, a three-party call is a relationship between terminals, and a call forwarding variable is a relationship between terminals.

A *terminal action* is shown in Figure 9.19. Here's one interpretation of that semantic network:

Change to disconnected PART_OF change in line state.
Change to conversing PART_OF change in line state.
Change to on-hold PART_OF change in line state.

Here's another interpretation:

Change to disconnected IS_A change in line state.
Change to conversing IS_A change in line state.
Change to on-hold IS_A change in line state.

From Figure 10.5, the following interpretations are possible:

Handle PART_OF automobile.
Body PART_OF automobile.
Tire PART_OF automobile.
Brake PART_OF automobile.

From Figure 10.5, the following interpretations are not possible:

Handle IS_A automobile.
Brake IS_A automobile.

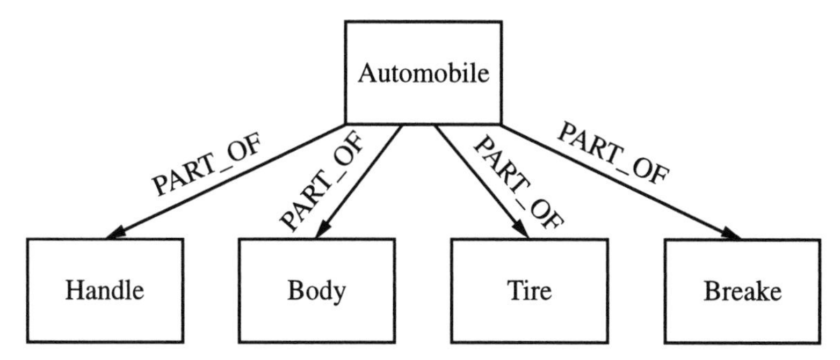

FIGURE 10.5 Semantic network about an automobile.

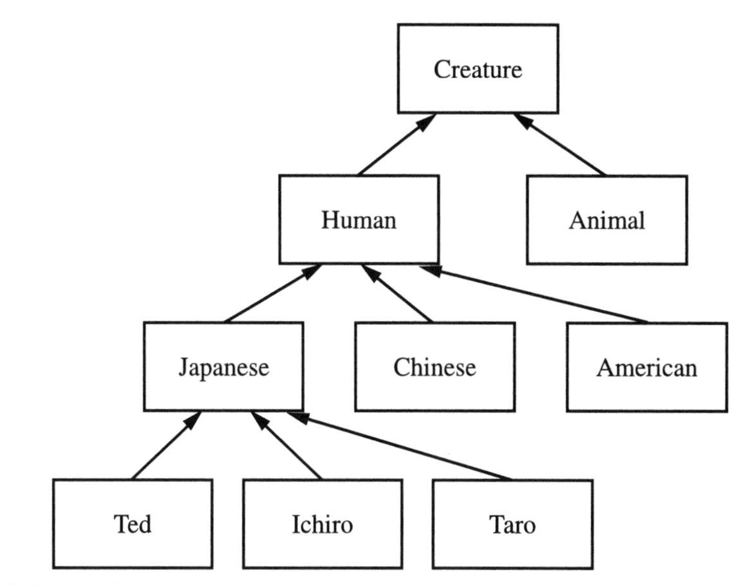

FIGURE 10.6 Creature network.

From Figure 10.6, the following interpretations are possible:

Ted IS_A Japanese.
Japanese IS_A human.
Human IS_A creature.

But another interpretation is possible:

Ted PART_OF Japanese.
Japanese PART_OF human.
Human PART_OF creature.

Generally speaking, when A IS_A B, A is a concrete object and B is an abstract concept that includes the concrete object A. In other words, B is a common concept underlying object A. For example, the common concept underlying a Japanese, a Chinese, and an American is a human. The common concept underlying a human and an animal is a creature. When A PART_OF B, A is a component of B. For example, a Japanese may be a component of a human. A Chinese is also a component of a human.

10.3 SYMBOLIC LOGIC

Predicate logic is used for telecommunication services such as intelligent directory service and intelligent switching service. In this section, predicate logic as well as propositional logic are described.

Mathematical logic includes predicate logic and propositional logic. Each of the following two sentences is a *proposition*, that is, a description of a specific domain:

(1) The earth goes around the sun.
(2) A telephone is a terminal for telecommunication services.

Each proposition is composed of a primitive proposition or a compound proposition, which is a combination of primitive propositions.

Every proposition is either true (T) or false (F). Proposition (1) and (2) are true. On the other hand, the propositions "The earth is a triangle" and "A human is a bird" are false.

10.3.1 Definition of a Proposition and Its Operations

A *proposition* can be represented by a symbol, called an *atomic formula*, such as p, q, r, and s. An atomic formula is either true (T) or false (F).

An operator that performs an operation on one or more atomic formulas is called a *logical symbol*, for example, $\leftrightarrow$, $\sim$, $\rightarrow$, $\vee$, or $\wedge$. The priority of operators is as follows:

(1) $\sim$
(2) $\wedge$
(3) $\vee$
(4) $\rightarrow$
(5) $\leftrightarrow$

where (1) is the highest priority and (5) is the lowest priority. The operators have the following meanings:

"$\sim$" is a negation.
"$\wedge$" is a conjunction.

TABLE 10.1 Logical Operation

p	q	$p \wedge q$	$p \vee q$	$\sim q$	$p \leftrightarrow q$	$p \rightarrow q$
T	T	T	T	F	T	T
T	F	F	T	T	F	F
F	T	F	T	F	F	T
F	F	F	F	T	T	T

TABLE 10.2 Operation of $\sim p \vee q$

p	q	$\sim p \vee q$
T	T	T
T	F	F
F	T	T
F	F	T

"$\vee$" is a disjunction.
"$\rightarrow$" is an implication.
"$\leftrightarrow$" is an equivalent.

The definitions of these operators are shown in Table 10.1.

Table 10.2 shows that the operation "$\sim p \vee q$" is equivalent to the operation "$p \rightarrow q$."

The following are true of a logical expression.

(1) Both an atomic formula and a negation of an atomic formula are logical expressions.
(2) A formula of logical expressions connected by an operator or operators is a logical expression.
(3) Only a formula defined as in items (1) and (2) is a logical expression.

The operation rules governing logical expressions are as follows:

(1) $p \leftrightarrow q = (p \rightarrow q) \wedge (p \rightarrow p)$
(2) $p \rightarrow q = \sim p \vee q$
(3) $p \wedge \mathrm{T} = p$
(4) $p \wedge \mathrm{F} = \mathrm{F}$
(5) $p \vee \mathrm{T} = \mathrm{T}$
(6) $p \vee \mathrm{F} = p$
(7) $\sim(\sim p) = p$
(8) $p \wedge p = p$
(9) $p \vee p = p$
(10) $p \vee \sim p = \mathrm{T}$
(11) $p \wedge \sim p = \mathrm{F}$

(12) $p \wedge q = q \wedge p$
(13) $p \vee q = q \vee p$
(14) $p \wedge (q \vee r) = (p \wedge q) \vee (p \wedge r)$
(15) $p \vee (q \wedge r) = (p \vee q) \wedge (p \vee r)$
(16) $\sim(p \wedge q) = \sim p \vee \sim q$
(17) $\sim(p \vee q) = \sim p \wedge \sim q$
(18) $(p \wedge q) \wedge r = p \wedge (q \wedge r)$
(19) $(p \vee q) \vee r = p \vee (q \vee r)$

10.3.2 Clausal Form

A *literal* is an atomic formula or a negation of an atomic formula. A disjunction of a literal is considered a clause C_i, which is notated as follows:

$$C_i\ (i = 1, 2, \cdots, n) = P_{i1} \vee P_{i2} \vee \cdots \vee P_{in}$$

A clausal form C is a conjunction of clauses, described as follows:

$$C = C_1 \wedge C_2 \wedge \cdots \wedge C_n$$

A logical expression is converted to a clausal form via the following operations.

(1) Eliminate an equivalent and an implication from a logical expression:

$$p \leftrightarrow q = (p \rightarrow q) \wedge (q \rightarrow p)$$

$$p \rightarrow q = \sim p \vee q$$

(2) Move a negation to the front of an atomic formula:

$$\sim(\sim p) = p$$

$$\sim(p \wedge q) = \sim p \vee \sim q$$

$$\sim(p \vee q) = \sim p \wedge \sim q$$

(3) Apply a distribution rule to a logical expression:

$$p \wedge (q \vee r) = (p \wedge q) \vee (p \wedge r)$$

$$p \vee (q \wedge r) = (p \vee q) \wedge (p \vee r)$$

Here are some examples of how a clausal form is obtained from a logical expression by the operations just described.

Example 1:

$$\begin{aligned}(p \rightarrow q) \vee (r \rightarrow s) \wedge (t \rightarrow u) &= (\sim p \vee q) \vee (\sim r \vee s) \wedge (\sim t \vee u)\\ &= \{(\sim p \vee q) \vee (\sim r \vee s)\} \wedge \{(\sim p \vee q) \vee (\sim t \vee u)\}\\ &= (\sim p \vee q \vee \sim r \vee s) \wedge (\sim p \vee q \vee \sim t \vee u)\end{aligned}$$

Example 2:

$$(p \to q) \vee (r \wedge s) \vee (t \wedge u) = (\sim p \vee q \vee r) \wedge (\sim p \vee q \vee s) \vee (t \wedge u)$$
$$= (t \wedge u) \vee (\sim p \vee q \vee r) \wedge (\sim p \vee q \vee s)$$
$$= \{(t \wedge u) \vee (\sim p \vee q \vee r)\} \wedge \{(t \wedge u) \vee (\sim p \vee q \vee s)\}$$
$$= \{(\sim p \vee q \vee r \vee t)\} \wedge (\sim p \vee q \vee r \vee u)\} \wedge$$
$$\{(\sim p \vee q \vee s \vee t)\} \wedge (\sim p \vee q \vee s \vee u)\}$$

Example 3:

$$(p \wedge q) \vee (r \wedge s) \vee (t \wedge u)$$

Example 4:

$$(p \wedge q) \vee (r \to s) \vee (t \to u)$$

A logical expression is true, false, or true or false. A logical expression with value "T" is called a tautology and is valid. A logical expression with value "F" is a contradiction and is unsatisfactory. A logical expression of value "T" or "F" is satisfactory.

10.3.3 Evaluation of a Logical Expression by Means of a Semantic Tree

Consider the logical expression P composed of atomic formulas p_1, p_2, p_n. P can be represented as $P(p_1, p_2, \ldots, p_n)$.

An evaluation method in which both "T" and "F" are assigned to each atomic formula is the semantic tree method:

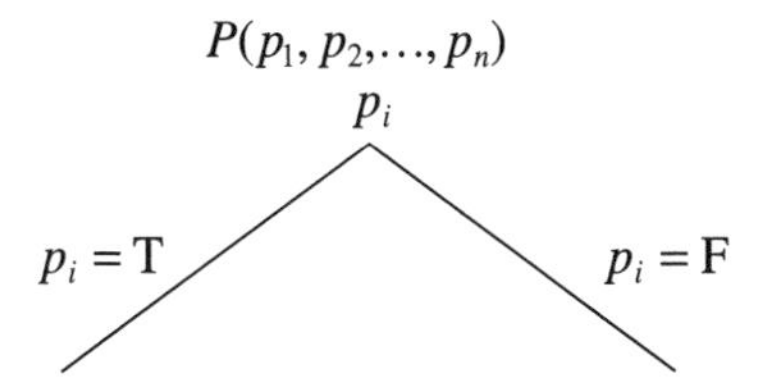

In this method, an evaluation proceeds in finite steps by assigning "T" and "F" to each atomic formula. Some examples follow.

Example 1:

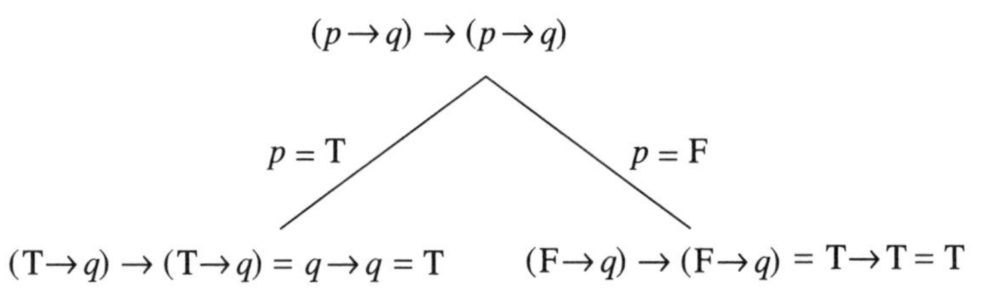

This expression is a tautology and is valid.

Example 2:

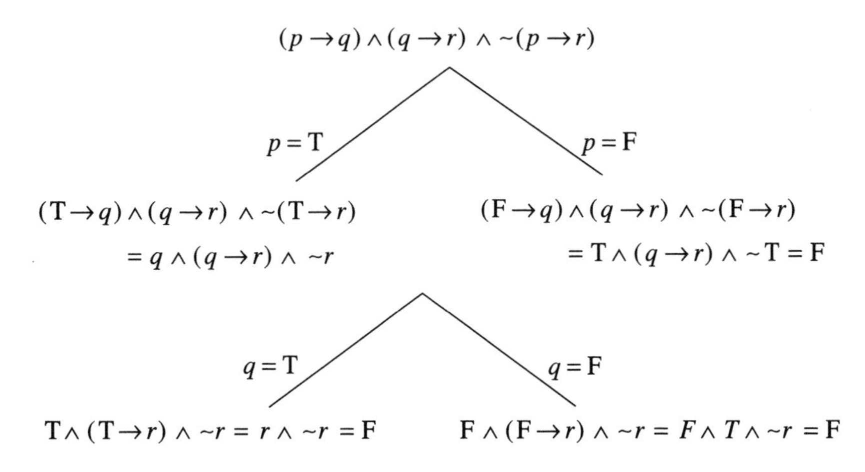

This expression is a contradiction and is unsatisfactory.

Example 3:

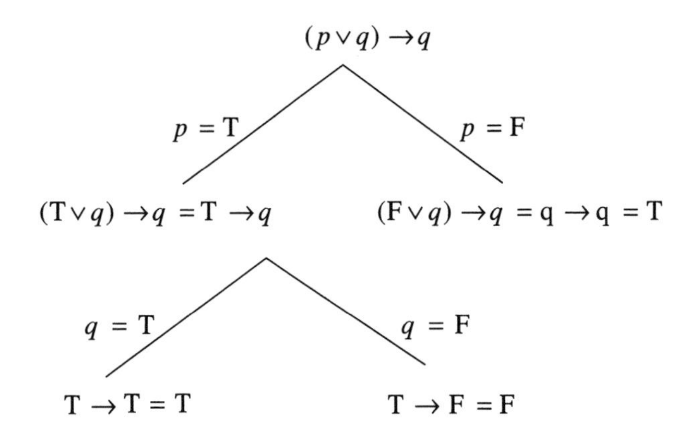

The expression is satisfactory.

10.4 PREDICATE LOGIC

Propositional logic deals with a statement in a logical function, whereas predicate logic uses semantics and an internal structure of a sentence. For example, given the two propositions, "A human dies" and "Ted is a human," we can infer "Ted dies." With propositional logic this problem is solved as follows: If "A human dies" is represented as p, "Ted is a human" is represented as q, and "Ted dies" is represented as r, then $(p \wedge q) \to r$ should have the value T. This interpretation is performed as follows:

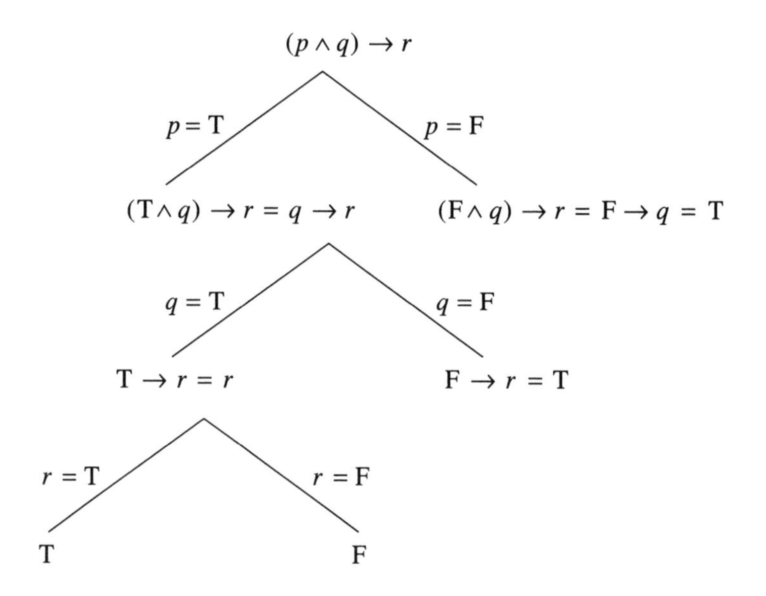

According to propositional logic, $(p \wedge q) \to r$ is not valid but is satisfactory. So this problem cannot be solved by propositional logic. Instead, it should be solved as follows:

"A human dies" is notated

$$\text{Human}(x) \to \text{Die}(x) \tag{1}$$

"Ted is a human" is notated

$$\text{Human(Ted)} \tag{2}$$

By assigning "Ted" to x in statement (1), we obtain Die(Ted).

Human and Die are called *predicates*. Ted and x are called *parameters*. In addition to predicate symbols such as Human, Die, and Ted, predicate logic introduces a universal quantifier ($\forall$) and an existential quantifier ($\exists$). Using these quantifiers, statement (1) can be notated as:

$$\forall x[\text{Human}(x) \to \text{Die}(x)] \tag{3}$$

This means that if x is a human, then x dies for all x.

An existential quantifier is used as follows:

$$\exists x[\text{Flower}(x) \to \text{Red}(x)] \tag{4}$$

This means that if x is a flower, then there is a red flower x.

10.4.1 Definitions and Operations for Predicate Logic

An *individual constant* represents a specific object and is notated a, b, c,....

An *individual variable* represents any object and notated x, y, z,....

A *functional symbol* represents a relation between or among objects and is notated $f(x, y)$, $g(z, w)$,.... Here the functional symbol g shows the relationship between z and w.

A *predicate symbol* represents a predicate for objects and is notated $P(x, y)$, $Q(z)$,..., where P and Q are predicate symbols.

A *logical symbol* represents an operation on predicate symbols and is notated $\leftrightarrow$, ~, $\rightarrow$, $\vee$, or $\wedge$.

A *term* can contain individual constants, individual variables, and/or functions.

Quantifiers come in two forms: existential quantifier ($\exists$) and universal quantifier ($\forall$). For example, our earlier statement (3) means that if x is a human, then x dies for all x. And our earlier statement (4) means that a red flower x exists if x is a flower.

In predicate logic a logical expression is defined as follows:

(1) If $t_1, t_2,\ldots, t_n$ are terms and P is a predicate with n parameters, then $P(t_1, t_2,\ldots, t_n)$ is an atomic formula and a logical expression.

(2) If $P(t_1, t_2,\ldots, t_n)$ and $Q(s_1, s_2,\ldots, s_m)$ are logical expressions, then $\sim P(t_1, t_2,\ldots, t_n)$, $P(t_1, t_2,\ldots, t_n) \wedge Q(s_1, s_2,\ldots, s_m)$, $P(t_1, t_2,\ldots, t_n) \vee Q(s_1, s_2,\ldots, s_m)$, $P(t_1, t_2,\ldots, t_n) \rightarrow Q(s_1, s_2,\ldots, s_m)$, $Q(s_1, s_2,\ldots, s_m)$, and $P(t_1, t_2,\ldots, t_n) \leftrightarrow Q(s_1, s_2,\ldots, s_m)$ are logical expressions.

(3) If $P(x_1, x_2,\ldots, x_n)$ is a logical expression, then $\forall x_1, \forall x_2,\ldots, \forall x_n\, P(x_1, x_2,\ldots, x_n)$ and $\exists x_1, \exists x_2,\ldots, \exists x_n\, P(x_1, x_2,\ldots, x_n)$ are logical expressions.

A variable qualified by $\forall$ or $\exists$ is a *bound* variable. A variable that is not qualified by $\forall$ or $\exists$ is a *free* variable. A logical expression operation of predicate logic is shown in Table 10.3.

An area that a bound variable influences is called a *scope of the variable.* When a quantifier q is given, the scope of $qx[\ldots]$ is $[\ldots]$ for x, and the scope of $qx1[\ldots qx2[\ldots]]$ is $[\ldots qx2[\ldots]]$ for $x1$ and $[\ldots]$ for $x2$. Interpretations of $\forall x \exists y P(x, y)$ and $\forall x \exists y P(x, y)$ are different and shown in Figures 10.7 and 10.8, respectively, where x is one of the elements $\{a, b, c\}$and y is one of the elements $\{d, e, f\}$.

TABLE 10.3 Logical Operation of Predicate Logic

P∧Q	P∨Q	P→Q	P↔Q	~P	~Q
T	T	T	T	F	F
F	T	F	F	F	T
F	T	T	F	T	F
F	F	T	T	T	T

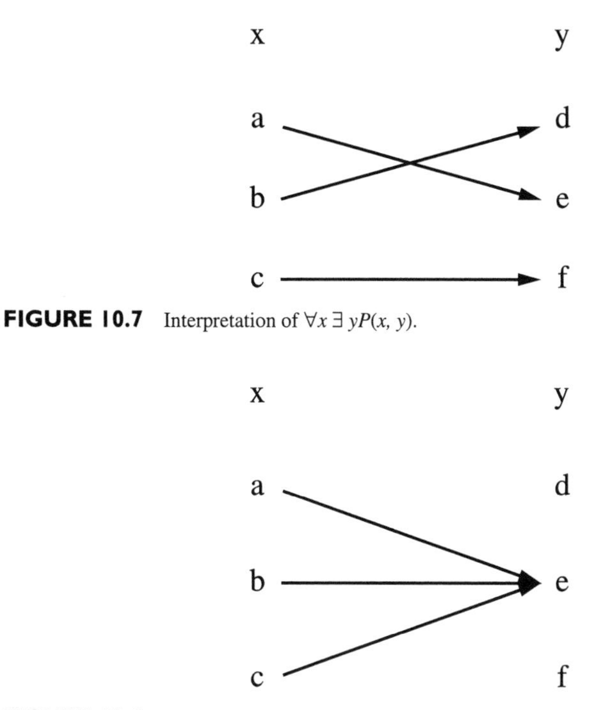

FIGURE 10.7 Interpretation of $\forall x \exists yP(x, y)$.

FIGURE 10.8 Interpretation of $\exists y \forall xP(x, y)$.

10.4.2 Clausal Form

An atomic formula or a negation of an atomic formula is called a *literal*. If a clause $C_i (i = 1, 2, \ldots, m)$ is composed of a disjunction of literals, then $q_{x1}, q_{x2}, \ldots, q_{xn} (C_1 \wedge C_2 \wedge \cdots \wedge C_m)$ is called a *clausal form*, where q is a quantifier $\forall$ or $\exists$ and $x_1, x_2, \ldots, x_n$ are bound variables that appear in $C_1 \wedge C_2 \wedge \cdots \wedge C_m$. $C_1 \wedge C_2 \wedge \cdots \wedge C_m$ is called a *matrix*.

Conversion from a logical expression to a clausal form is accomplished as follows.

(1) Eliminate an equivalent or an implication:

$$P \leftrightarrow Q = (P \rightarrow Q) \wedge (Q \rightarrow P)$$

$$(P \rightarrow Q) = \sim P \vee Q$$

(2) Move a negation to the front of an atomic formula:

$$\sim(\sim P) = P$$

$$\sim(P \wedge Q) = \sim P \vee \sim Q$$

$$\sim(P \vee Q) = \sim P \wedge \sim Q$$

$$\sim\forall xP(x) = \exists x[\sim P(x)]$$

$$\sim\exists xP(x) = \forall x[\sim P(x)]$$

(3) Rename bound variables so they do not conflict with one another:

$$\forall xP(x) = \forall yP(y)$$

$$\exists xP(x) = \exists yP(y)$$

(4) Convert form (3) to a standard form:

$$\forall xP(x) \wedge Q = \forall x[P(x) \wedge Q]$$

$$\forall xP(x) \vee Q = \forall x[P(x) \vee Q]$$

$$\exists xP(x) \wedge Q = \exists x[P(x) \wedge Q]$$

$$\exists xP(x) \vee Q = \exists x[P(x) \vee Q]$$

(5) Convert a matrix to a conjunctive normal form:

$$\begin{aligned}\forall x \exists yP(x, y) \rightarrow \exists x \forall yQ(x, y) &= \sim(\forall x \exists yP(x, y)) \vee \exists x \forall yQ(x, y)\\ &= \exists x \forall y(\sim P(x, y)) \vee \exists x \forall yQ(x, y)\\ &= \exists x_1 \forall y_1(\sim P(x_1, y_1)) \vee \exists x_2 \forall y_2Q(x_2, y_2)\\ &= \exists x_1 \forall y_1\exists x_2 \forall y_2[\sim P(x_1, y_1) \vee Q(x_2, y_2)]\end{aligned}$$

A predicate logical expression is *satisfactory* when it has at least a value T. A logical expression is *valid* when it always has a value T. A logical expression is *unsatisfactory* when it always has a value F.

10.4.3 Herbrand Universe and the Herbrand Theorem

To prove that a clausal set is unsatisfactory, it is necessary to consider every interpretation in a specific domain. However, this is difficult to achieve. Instead, a Herbrand universe that is equivalent to the above interpretation is considered.

A Herbrand universe is defined as follows.

(1) H_0 = {a set of individual constants in a clause set C}. Unless there are no individual constants, it is assumed that $H_0 = \{a\}$.
(2) $H_{i+1} = H_i \cup$ {a set of $f(s_1, s_2, s_n)$}, where f is all of the functions used in C and S_i is every ground instance in H_i.
(3) $H(C) = H_\infty$

Example 1: $C = \{P(a), Q(x)\}$, so $H_0 = \{a\}$.

Example 2: $C = \{P(a), Q(f(x)), R(x)\}$, so $H_0 = \{a\}$, $H_1 = \{a, f(a)\}$, and $H_2 = \{a, f(a), f(f(a))\}$.

An atomic formula of a clause set C, where an element of $H(C)$ is assigned to a variable of the atomic formula, is called a *ground instance*. A set composed of all of the ground instances is called a *Herbrand base*. Because an atomic formula

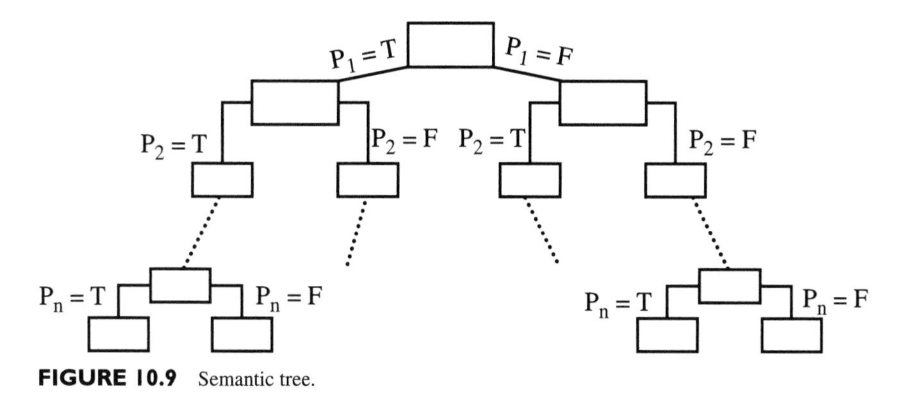

FIGURE 10.9 Semantic tree.

in a Herbrand base does not have any variables, it is possible to assign a value, T or F, to the atomic formula. If a Herbrand base is composed of n variables, then 2^n interpretations are possible.

Example 3: The Herbrand base of Example 1 is $\{P(a), Q(a)\}$. The Herbrand base of Example 2 is $\{P(a), Q(f(a)), R(a), Q(f(f(a))), R(f(a)), \ldots\}$. When a Herbrand base $\{P(a), Q(a)\}$ of a clause set C is given, the interpretation is as shown in Figure 10.9. If C is a contradiction, there exists a node whose value is F on every branch of a semantic tree. The node whose value is F and appears first is called a *failure node*. If a clause set C is unsatisfactory, there exists a clause set C_1 that is composed of nodes whose values are false for all branches on a semantic tree. On the other hand, if there exists a clause set C that is unsatisfactory, then C is unsatisfactory. This is the *Herbrand theorem*.

Example 4: A given clause set $C = \{\sim P(a) \vee Q(b), \sim Q(x) \vee R(y), P(z), \sim R(w)\}$ has a Herbrand universe $H(C) = \{a, b\}$ and Herbrand base $HB(C) = \{P(a), P(b), Q(b), Q(a), R(a), R(b)\}$. Using the ground instance, the semantic tree in Figure 10.10 is generated. If there is a failure node at every branch of a semantic tree, then the clause set is unsatisfactory. That a clause set C_1 is unsatisfactory is proven by the following operations.

$$
\begin{aligned}
C_1 &= \{\sim P(a) \vee Q(b), \sim Q(b) \vee R(a), P(a), \sim R(a)\} \\
&= (\sim P(a) \vee Q(b)) \wedge (\sim Q(b) \vee R(a)) \wedge P(a) \wedge \sim R(a) \\
&= \{P(a) \wedge (\sim P(a) \vee Q(b))\} \wedge \{\sim R(a) \wedge (\sim Q(b) \vee R(a))\} \\
&= \{(P(a) \wedge \sim P(a)) \vee (P(a) \wedge Q(b))\} \wedge \{(\sim R(a) \wedge \sim Q(b)) \vee (\sim R(a) \wedge R(a))\} \\
&= \{F \vee (P(a) \wedge Q(b))\} \wedge \{(\sim R(a) \wedge \sim Q(b) \vee F\} \\
&= (P(a) \wedge Q(b)) \wedge (\sim R(a) \wedge \sim Q(b)) \\
&= P(a) \wedge \sim R(a) \wedge Q(b) \wedge \sim Q(b) \\
&= P(a) \wedge \sim R(a) \wedge F \\
&= F
\end{aligned}
$$

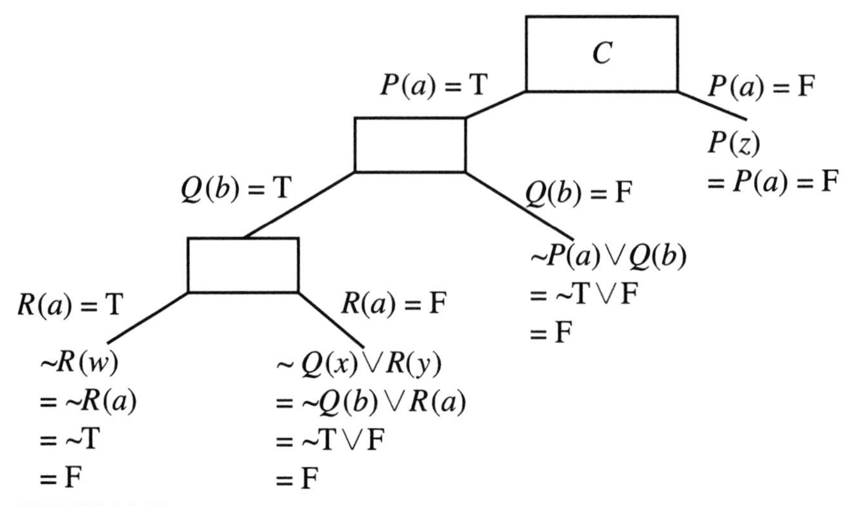

FIGURE 10.10 Semantic tree of $\{\sim P(a) \vee Q(b), \sim Q(x) \vee R(y), P(z), \sim R(w)\}$.

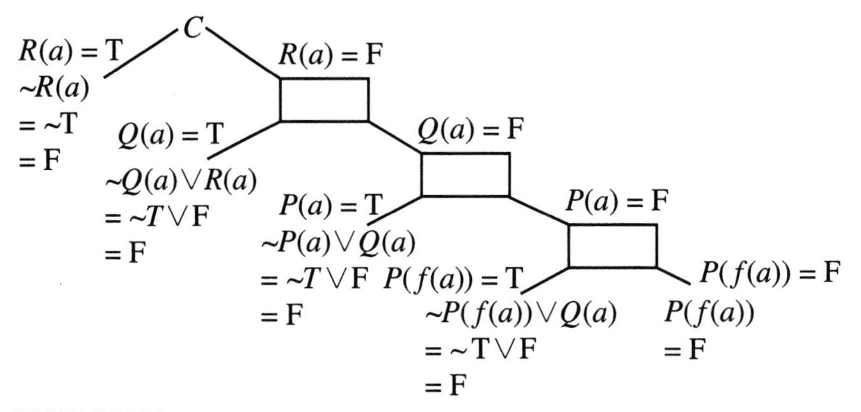

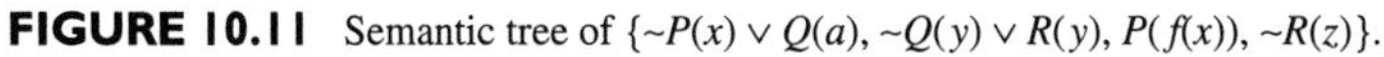
FIGURE 10.11 Semantic tree of $\{\sim P(x) \vee Q(a), \sim Q(y) \vee R(y), P(f(x)), \sim R(z)\}$.

Example 5: The following clause set is given:

$$C = \{\sim P(x) \vee Q(a), \sim Q(y) \vee R(y), P(f(x)), \sim R(z)\}$$

The Herbrand universe and Herbrand base are as follows:

$$H(C) = \{a, f(a), f(f(a)), f(f(a)), \ldots\},$$
$$HB(C) = \{P(a), Q(a), R(a), P(f(a)), Q(f(a)), R(f(a)), \ldots\}$$

A semantic tree is shown in Figure 10.11. Because the tree has failure nodes on all of the branches, a clause set C is unsatisfactory. A set of clauses with ground instances is defined as C' and notated as follows:

$$C' = \{\sim P(a) \vee Q(a), \sim P(f(a)) \vee Q(a), \sim Q(a) \vee R(a), P(f(a)), \sim R(a)\}$$

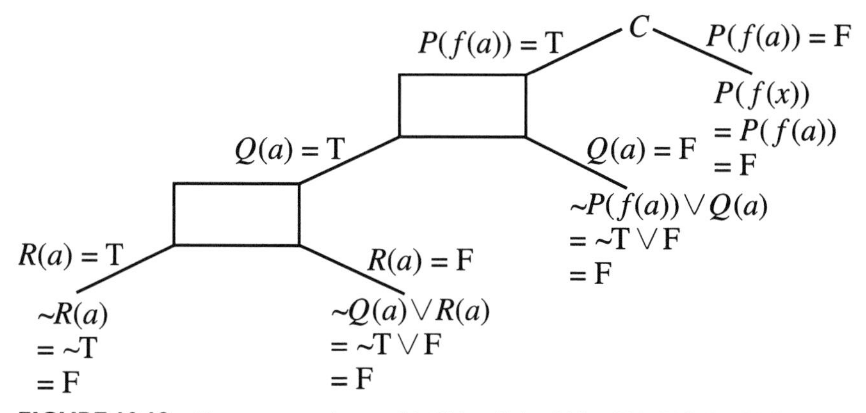

FIGURE 10.12 Alternate semantic tree of $\{\sim P(x) \vee Q(a), \sim Q(y) \vee R(y), P(f(x)), \sim R(z)\}$.

Let's find another clausal form. In Example 4, if it is assumed that $x = y = z = a$, then the semantic network in Figure 10.12 is obtained. A set of clause sets with ground instances is defined as C″ and notated as follows:

$$C'' = \{\sim P(f(a)) \vee Q(a), \sim Q(a) \vee R(a), P(f(a)), \sim R(a)\}$$

Thus, there is more than one clausal form with ground instances that are unsatisfactory. To prove that C is unsatisfactory, do the following:

(1) Make sets of clauses with ground instances C' C'', ...
(2) Prove one of the sets is unsatisfactory.

10.4.4 Proof of Tautology

To prove a tautology of a clausal form

$$q_{x1}\, q_{x2} \cdots q_{xn}\, [C_1 \wedge C_2 \wedge \cdots \wedge C_m] \qquad (1)$$

a contradiction of the clausal form is proved by making a negation of either C_1, C_2,... or C_m. If it is proved, the clausal form is a tautology. For example, by negating C_i, the clausal form is described as follows:

$$\begin{aligned} & q_{x1}\, q_{x2} \cdots q_{xi} \cdots q_{xn}\, [C_1 \wedge C_2 \wedge \cdots \wedge \sim C_i \wedge \cdots \wedge C_m] \\ & = q_{x1}\, q_{x2} \cdots q_{xi} \cdots q_{xn}\, [\sim (C_1 \wedge C_2 \wedge \cdots \wedge C_i \wedge \cdots \wedge C_m) \qquad (2) \end{aligned}$$

If clausal form (2) is a contradiction, then clausal form (1) is a tautology.

There is a universal quantifier ∀ and an existential quantifier ∃ in logical expressions. Consider the following example:

$$\forall x_1\, \forall x_2\, \forall x_3\, \exists x_4\, P(x_1\, x_2\, x_3\, x_4) \qquad (3)$$

where x_4 is dependent on x_1, x_2 and x_3 and the relationship among variables x_1, x_2, x_3, and x_4 is as follows:

$$x_4 = f(x_1, x_2, x_3)$$

Then expression (3) is represented as follows:

$$\forall x_1 \forall x_2 \forall x_3\ P(x_1, x_2, x_3, f(x_1, x_2, x_3)) \tag{4}$$

Thus, the existential quantifier $\exists$ is eliminated. $f(x_1, x_2, x_3)$ is called a *Skolem function*. Expression (4) is called a *Skolem standard form*. In this way the clausal form without existential quantifiers is obtained. In expression (4), $P(x_1, x_2, x_3, f(x_1, x_2, x_3)$ is called a *clause set.*

10.4.5 Resolution Principle

In a clause set C, an operation that derives C_{ij} from C_i and C_j by excluding P and $\sim P$ is called a *resolution*, where C_i has an atomic formula P and C_j has atomic formula $\sim P$. Adding C_{ij} to C, a new clausal set $C \cup \{C_{ij}\}$ is created. This operation is performed repeatedly until a null clause is derived. If a null clause is derived, the clause set is unsatisfactory. This operation is called a *resolution principle.* When a resolution principle is applied to $C_1 = \sim P \vee Q(1)$ and $C_2 = \sim Q \vee R(2)$, then a new clause $C_{12} = \sim P \vee R$ (3) is obtained.

A resolution principle has the same effect as a *syllogism.* According to the syllogism, a new logical expression $P \rightarrow R$ is derived from the expression $P \rightarrow Q$ and the expression $Q \rightarrow R$:

$$P \rightarrow Q = \sim P \vee Q \tag{5}$$

$$Q \rightarrow R = \sim Q \vee R \tag{6}$$

$$P \rightarrow R = \sim P \vee R \tag{7}$$

Through the resolution principle, Eq. (7) is obtained from Eqs. (5) and (6). Therefore a syllogism is the same as the resolution principle.

Example 1: A clause set $C_1 = \{\sim P(a) \vee Q(b), \sim Q(x) \vee R(y), P(z), \sim R(t)\}$ is given. The resolution principle is applied as in Figure 10.13, where u is a unification and $u\{b/x\}$ means that b is assigned to x.

10.4.6 Logical Consequence

(Definition) Consider the logical expression $(P_1, P_2, \ldots, P_n)$ and the logical expression Q. If $(P_1, P_2, \ldots, P_n)$ is assumed to be true and Q becomes true, then Q is a logical consequence of $(P_1, P_2, \ldots, P_n)$. If and only if a logical expression $(P_1, P_2, \ldots, P_n) \rightarrow Q$ is valid, then $(P_1, P_2, \ldots, P_n) \rightarrow Q$ is true. Then Q is a logical consequence of $(P_1, P_2, \ldots, P_n)$. (Proof) Eq. $(P_1, P_2, \ldots, P_n) \rightarrow Q$ is assumed to be valid. Then if $(P_1, P_2, \ldots, P_n)$ is true, Q is true. According to this definition, Q is a logical consequence of $(P_1, P_2, \ldots, P_n)$, On the other hand, if Q is a logical consequence of $(P_1, P_2, \ldots, P_n)$, Q is true when $(P_1, P_2, \ldots, P_n)$ is true. Therefore

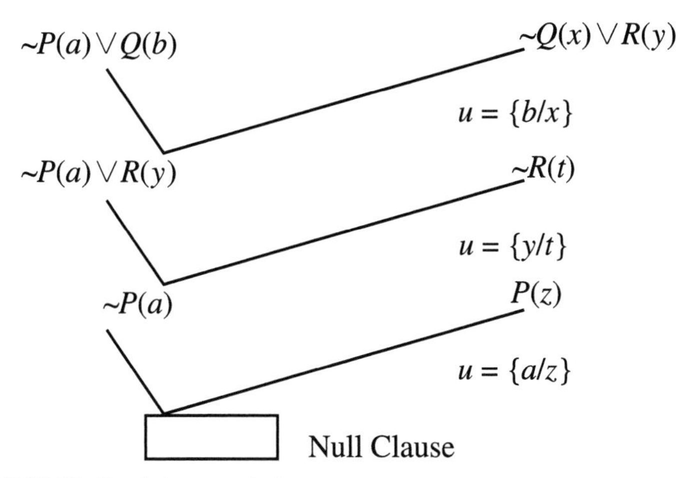

FIGURE 10.13 Resolution example 1.

$(P_1, P_2,\ldots, P_n) \rightarrow Q = T \rightarrow T = T$ is proven and is valid. If $(P_1, P_2,\ldots, P_n)$ is false, $(P_1, P_2,\ldots, P_n) \rightarrow Q$ is true, no matter what the value of Q is. Therefore, if and only if $(P_1, P_2,\ldots, P_n) \rightarrow Q$ is valid, then Q is a logical consequence of $(P_1, P_2,\ldots, P_n)$.

Example: To prove that Q is a logical consequence of $(P_1 \wedge P_2 \wedge\cdot\cdot\wedge P_n)$, it must be proved that $(P_1 \wedge P_2 \wedge \wedge \wedge P_n) \rightarrow \text{Q}$ is true. Instead of proving this expression, it is enough to prove that $\sim\{(P_1 \wedge P_2 \wedge\cdot\cdot\wedge P_n) \rightarrow Q\}$ is false. $\sim\{(P_1 \wedge P_2 \wedge\cdot\cdot\wedge P_n) \rightarrow \text{Q}\}$ is transformed to $(P_1 \wedge P_2 \wedge\cdot\cdot\wedge P_n \wedge \sim Q)$. If $(P_1 \wedge P_2 \wedge\cdot\cdot\wedge P_n \wedge \sim Q)$ is proved to be false, then $(P_1, \wedge P_2 \wedge\cdot\cdot\wedge P_n) \rightarrow Q$ is true.

10.4.7 Horn Set

A clause set C is a conjunction of $C_1, C_2, \ldots, C_n$. A clause is a disjunction of literals, where a literal is an atomic formula or a negation of an atomic formula. Then a clause C is represented as follows:

$$C = \sim P \vee \sim P_2 \vee\cdots\vee\sim P_n \vee Q_1 \vee Q_2 \vee\cdots\vee Q_m. \tag{8}$$

Equation (8) is transformed as follows:

$$\begin{aligned} \text{C} &= \sim(P_1 \wedge P_2 \wedge\cdots\wedge P_n) \vee (Q_1 \vee Q_2 \vee\cdots\vee Q_m) \\ &= (P_1 \wedge P_2 \wedge\cdots\wedge P_n) \rightarrow (Q_1 \vee Q_2 \vee\cdots\vee Q_m) \end{aligned} \tag{9}$$

Considering the right side of Equation (9), which consists of only one atomic formula, Equation (9) is shown as

$$(P_1 \wedge P_2 \wedge\cdots\wedge P_n) \rightarrow Q \tag{10}$$

Equation (10) is represented as follows:

$$Q \leftarrow P_1, P_2,\ldots, P_n \tag{11}$$

This means that if $(P_1, P_2,\ldots, P_n)$ is true, then Q is true. Equation (11) is called a *Horn clause*. The following expressions are defined based on a Horn clause:

$$Q \leftarrow P_1, P_2,\ldots, P_n \tag{12}$$

$$Q \leftarrow \tag{13}$$

$$\leftarrow P_1, P_2,\ldots, P_n \tag{14}$$

$$\leftarrow \tag{15}$$

Equation (12) means that Q is true if $(P_1, P_2,\ldots, P_n)$ is true. Equation (13) means that Q is unconditionally true. Equation (14) means it is a question whether $(P_1, P_2,\ldots, P_n)$ is true or not. In other words, it is a goal clause. Equation (15) is a null clause.

A set composed of Eqs. (12), (13), (14), and (15) is called a Horn set. Using a Horn set, a resolution is performed as follows: Make a goal clause one of the parent clauses. Select a Horn clause expressed in Eq. (12) or Eq. (13) whose left literal is matched to any of the literals in a goal clause. The literal of the goal clause is replaced by the right part of the Horn clause (12) or (13). The operation continues until the result becomes a null clause. Here is an example of a resolution:

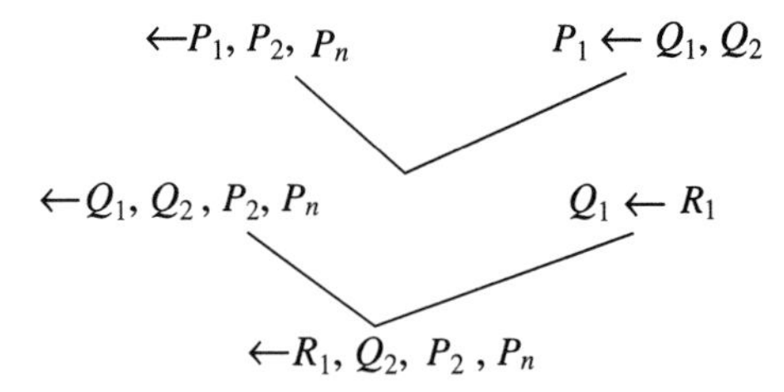

Example: The following Horn clauses are given.

Like$(x, y) \leftarrow$ Man(x), Beer(y)
Like$(x, y) \leftarrow$ Woman(x), Apple(y)
Man(Ted)
Beer(Asahi) $\leftarrow$
Woman(Hannah) $\leftarrow$
Apple(Katie) $\leftarrow$

When a goal clause $\leftarrow$ Like(Ted, Asahi) is given, The resolution in Figure 10.14 is conducted.

10.4.8 Application to Telecommunication Service

For example, consider directory service. The following Horn sets are defined to solve the goal clause:

$$S_e(S_0, S) \leftarrow S(S_0, S_1), V(S_1, S_2), O(S_2, S)$$

$$S([x1S], S) \leftarrow$$

$$V(S_0, S) \leftarrow AV(S_0, S_1), V(S_1, S)$$

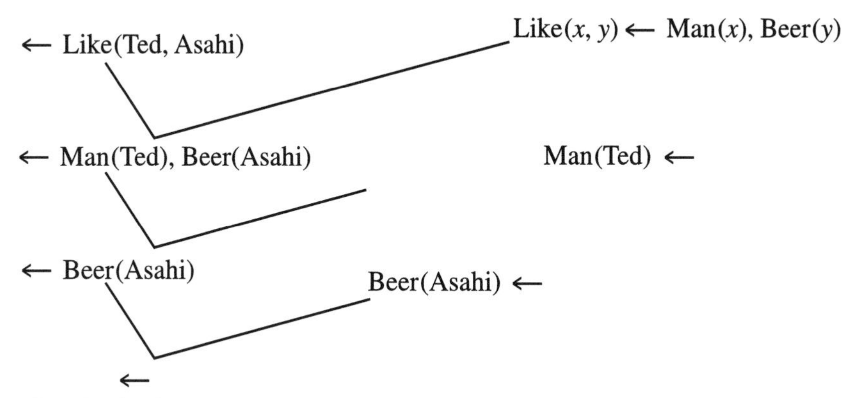

FIGURE 10.14 Resolution example.

$$AV([x, y, z \mid S], S) \leftarrow$$
$$V([x \mid S], S) \leftarrow$$
$$O(S_0, S) \leftarrow O_1(S_0, S_1), O_2(S_1, S_2), O_3(S_2, S)$$
$$O_1([x, y, z \mid S], S) \leftarrow$$
$$O_2([x, y, z, w \mid S], S) \leftarrow$$
$$O_3([x, y \mid S], S) \leftarrow$$

where S_0, S, S_1, and S_2 are lists. A list is described as $[x_1, x_2, x_n]$. $S_e(S_0, S)$ means that a sentence S_e is in the first part of the list S_0 and the remaining list is S. A query is issued as follows:

I would like to know the phone number of a clothing shop in San Francisco.

The sentence is transformed into a list [I, would, like, to, know, the, phone, number, of, a, clothing, shop, in, San Francisco]. A goal clause is represented as follows.

$$\leftarrow S_e(\text{[I, would, like, to, know, the, phone, number, of, a, clothing, shop, in, San Francisco]})$$

Then the resolution is performed. The results are obtained as follows (the resolution process is omitted here):

$$S([x \mid S], S) \leftarrow$$
$$[x] = [I]$$
$$AV(x, y, z \mid S, S) \leftarrow$$
$$[x, y, z] = \text{[would, like, to]}$$
$$N([x \mid S], S) \leftarrow$$

$$[x] = [\text{know}]$$

$$O_1([x, y, z \mid S], S) \leftarrow$$

$$[x, y, z] = [\text{the, phone, number}]$$

$$O_2([[x, y, z, w] \mid S], S) \leftarrow$$

$$[x, y, z, w] = [\text{of, a, clothing, shop}]$$

$$O_3([[x, y] \mid S], S) \leftarrow$$

$$[x, y] = [\text{in, San Francisco}]$$

In this way, the sentence is parsed and the word is detected in a list. After obtaining the word, a command is issued to the database to get the phone number of a clothing shop from the database.

11

TELESENSATION

Telesensation involves combining virtual reality (VR) with telecommunications to create telecommunication with realistic sensations. The term was coined by Dr. Nobuyoshi Terashima. Via Telesensation, an image of, say, a museum exhibit in a remote place is instantaneously transmitted through the communication links and displayed stereoscopically by using VR technology. Viewers can enter and walk through a scene that is a virtual world. Furthermore, they can even touch the leaves on a tree or a wall in a museum, as if they were physically there. This cuts the bonds of time and space.

Through Telesensation we can go anywhere without physically going there. We can enjoy sightseeing, go shopping, go to a museum, or go to school without leaving home. In this chapter, the concept and applications of Telesensation are described.

11.1 VIRTUAL REALITY CONCEPT

Virtual reality technology plays an important role in realizing Telesensation. Through it, a virtual world is created that viewers can enter and walk through and where they can handle virtual objects. The virtual world allows us a stereoscopic view from front or side, depending on our viewpoint, just as in the real world. The ability to enter and walk through the virtual world and handle virtual objects using hand gestures makes VR interactive, and this is one of its most important features.

Communication can be human–human communication, human–environment communication, or human–computer communication. In the case of human–human communication, a variety of means are at our disposal. We talk together to communicate. We write letters or draw pictures and sometimes communicate using images and motion pictures. In human–environment communication, we recognize our environment via our five senses: feeling, touch, taste, vision, and smell. In human–computer communication, we interact with a computer by means of a mouse, a touch pad, or a keyboard.

Human–human communication and human–environment communication have been developed over a long history of interaction. It is desirable to provide human beings with a human-friendly environment where we can interact with computers just as easily as we interact in human–human communication or human–environment communication.

The goal of VR is to provide human beings with a virtual environment where we can interact with a computer just as we do in the real world, that is, by talking with a virtual human in a spoken language, by writing a letter, or by drawing a picture. We can grasp a virtual object by hand gesture and bring it to another place. In a human-friendly virtual environment, we can interact with a computer without any difficulties or barriers. When a virtual landscape is generated by VR technology, we can go there just as if it were a real landscape. Providing not only a 3D image of the landscape but also sound and smell helps us enjoy the scenery.

In Oita prefecture, Japan, there is a museum where visitors can experience a virtual world. Upon entering the museum, we see a large screen in front of the seats. By sitting on a seat and wearing special glasses, visitors can enter a large virtual flower and smell it.

At ATR Communication Systems Laboratories, Kyoto, Japan, a virtual space teleconferencing system was developed in 1992. This next-generation video conference system provides participants a human-friendly environment for meeting and collaborating. They can view objects stereoscopically and have front or side views of the objects depending on their viewpoint. They can handle an object by means of hand gestures. In this system, each participant is at a different location, and all sites are connected via the network. Each site has a virtual conference room with a large screen in front of seats. On the screen, 3D images of real human beings are displayed stereoscopically, and participants can have stereoscopic views of various objects displayed on screen. They can have eye contact with each other. They can conduct a meeting as if they were gathered in the same place.

The image's motion is controlled by the real human's motion. The participant wears shutter glasses and has sensors on face, hands, and body to detect motion.

The shutter glasses provide a stereoscopic view. On the basis of the movement information, the images of the virtual person or object is deformed and displayed on the screen to match the viewer's perspective.

In the conventional video conference system, participants can meet face to face. However, it is very difficult to make eye contact and to have different views of objects according to the viewer's perspective. In the real world, a viewer can take a side view of an object just by moving to the side of the object. In the virtual space teleconferencing system, participants can make eye contact and take different views of the object to match their perspective. In the conventional video conference system, a participant cannot go inside the scene displayed on the screen, whereas in the virtual system he or she can enter the virtual space, walk through it, and grasp a virtual object by means of hand gesture, even feeling the heft of the object. To summarize, a viewer in the virtual world, can have a stereoscopic view of an object. This is called *stereoscopic display*. The viewer can enter the virtual world and walk through it. This is called *walk-through*. The viewer can take different views of the object according to his or her viewpoint. This is called *interaction*. The viewer can touch and grasp a virtual object and feel its heft. This is called *force feedback*. In the virtual world, even a collision can be detected.

To accomplish these functions, the viewer wears shutter glasses or a head-mounted display; this gives a stereoscopic view of the object, because the right image of the object enters the right eye and the left image enters the left eye. When the technologies of the lenticular screen and holography have been developed enough, viewers should be able to have a stereoscopic view with the naked eye. This involves detecting the viewer's viewpoint via a sensor attached near the viewer's eye and, based on this information, adjusting, the object's display on the screen. Object handling with hand gesture is accomplished by means of a data glove with sensor, that detects hand shape and hand motion. Again, this information is used to adjust display of the object. The technology for detecting hand motion and hand shape without use of a data glove is currently under study.

11.2 HISTORY OF VIRTUAL REALITY

Research into VR started in the 1960s. I.E. Sutherland developed a prototype of the head-mounted display. Dr. G.W. Furnes of Washington University developed a virtual flight simulator in the 1970s. An interactive art theater was developed in which a human being enters into a virtual scene as an actor or actress. Tele-existence research has also been conducted. A viewer enters a virtual space for training or operation of equipment as if he or she were there. The VPL company of the United States developed a system with a head-mounted display and a data glove for commercial purposes.

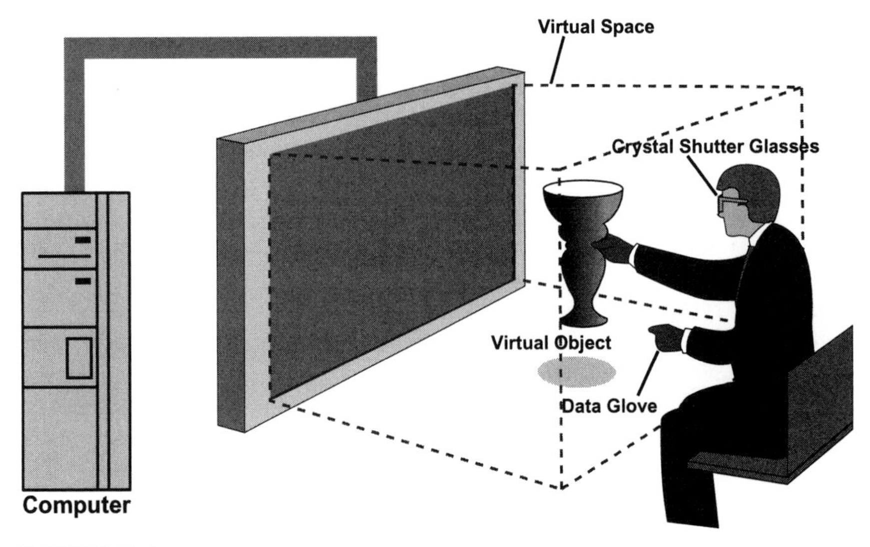

FIGURE 11.1 Virtual object handling system.

11.3 VIRTUAL OBJECT HANDLING

A schematic of a virtual object handling system is shown in Figure 11.1. In this system a virtual space is created in front of the screen in which a virtual object is displayed. A viewer in front of the screen can see the virtual object stereoscopically by wearing a pair of liquid crystal shutter glasses. Wearing a data glove he or she can handle the object. The data glove with sensors detects the shape, motion, and position of the hand. More specifically, it detects the 3D position of the hand. When the position of the hand and that of the object overlap, the system assumes that the hand is touching the object. The data glove enables the viewer to grasp or move the object just as in reality, touching and handling it directly with hand movements.

11.4 EXAMPLES OF VIRTUAL REALITY

The technology of VR creates a virtual space that inhabitants can enter and walk through and where they can touch objects just as in the real world. Many applications have already been developed. However, only some of them are discussed here: a walk-through system, a mountain bike system, a molecule visualization system, a modeling system, and a system for driving a streetcar.

In the walk-through system developed by North Carolina University, a viewer can enter a building, walk through it, go to a door and open it, and then go into the next room. In the mountain bike system, a viewer can ride a bike along the street or go through a mountain area. Special glasses present a front view. Turning to the right makes the scene on the right visible; turning to the left shows the left view. In the molecule visualization system, various kinds of molecules are shown in virtual space, and the viewer can enter the space and observe them. In the modeling system, a viewer can handle objects and lay them out in the virtual space. Using components of a city landscape, such as buildings, bridges, highways, and roads, he or she can construct an urban scene in virtual space. In the streetcar driving system, the viewer can drive a streetcar along a street while viewing the scene in front. The experience is realistic and enables the user to improve driving skill.

11.5 APPLICATIONS OF VIRTUAL REALITY

Medicine: VR can be used to simulate a medical operation and can provide tools for training.

Education: VR can be used for teaching a variety of subjects, such as geography, history, and chemistry. A viewer can experience a virtual museum scene or a virtual historical monument as if physically there. VR can reproduce architecture or ruins that have been lost and no longer exist on earth. In chemistry, students' understanding of molecular structures can be enhanced by visualizing and reproducing molecules in virtual space.

Telecommunication: The integration of VR and telecommunications results in Telesensation services, such as a virtual space teleconferencing system. With such a system, participants who are in reality at different sites have their images brought together through communication links so they can meet as if gathered in the same place. In a conventional video conferencing system, participants at remote locations—in reality their images—are brought through the communication links and displayed on screen so they can hold a meeting. However, their images are not displayed stereoscopically, they cannot have eye contact with one another, and they are unable to handle an object displayed on the screen. In the virtual space teleconferencing system, by comparison, participants' images are brought together in virtual space and displayed stereoscopically, allowing them eye contact as if they were gathered in the same room. This system was developed by the Advanced Telecommunication Research Institute (ATR). Through Telesensation, people at different sites can join together in virtual space and interact to carry out a job or simply to play. Telesensation can break the bonds of time and space.

Entertainment: Using VR, people can play a game with realistic sensations. A 3D virtual space is created where people participating in the game are displayed. A viewer can enter the space and enjoy games with the other participants.

Exhibition: A museum or an exhibition hall can be created and displayed stereoscopically. A viewer can enter it, walk through it, and appreciate paintings, statues, or pottery.

Architecture and civil engineering: The architecture of, say, a city hall or an art museum can be designed in 3D and displayed stereoscopically, allowing the viewer to see the architecture, enter it, and walk through it.

Science and academia: The flow of a fluid can be visualized and displayed stereoscopically. VR technology will contribute significantly to science and the academic world.

Aviation, space travel, and railway: Training for an operator can be performed in a virtual space.

11.6 TELESENSATION

Telesensation combines VR with telecommunications, allowing realistic sensations to be transmitted to destination. Through Telesensation, an image of a scene from a natural environment or images of humans at different sites are transmitted through communication links to a viewer and displayed stereoscopically on screen. The viewer can enter the scene, walk through it, and touch objects displayed there. Figure 11.2 shows a diagram of Telesensation. The scene of a street in Germany is taken by a camera and transmitted via broadband ISDN networks to Japan, where the scene is reproduced and displayed stereoscopically on screen by using VR. A viewer in Japan can enter the street scene and walk through it, can go to the entrance of a building, open the door, and go inside, and can even go behind the building and see what it looks like from there.

11.7 TYPES OF TELESENSATION

There are three main types of Telesensations: television-type service, interactive-type service, and teleconference service.

Television-type service: An image of a scene from a remote location is transmitted through communication links to a viewer and displayed stereoscopically on screen. The viewer can enter the scene, walk through it, and enjoy the sights. This type of service may use a 3D video theater or a video mural (see Figure 11.3).

Interactive-type service: An image of a scene from a remote place is taken by camera, transmitted via a communication link, and displayed stereoscopically in a virtual space to a viewer. The viewer can enjoy seeing the scene and walk through it, can touch or handle objects displayed in the virtual space, and can see the objects from different perspectives. This type of service will include teleshopping (see Figure 11.4), a tele-existence service, and telesimulation, such as a simulated golf game.

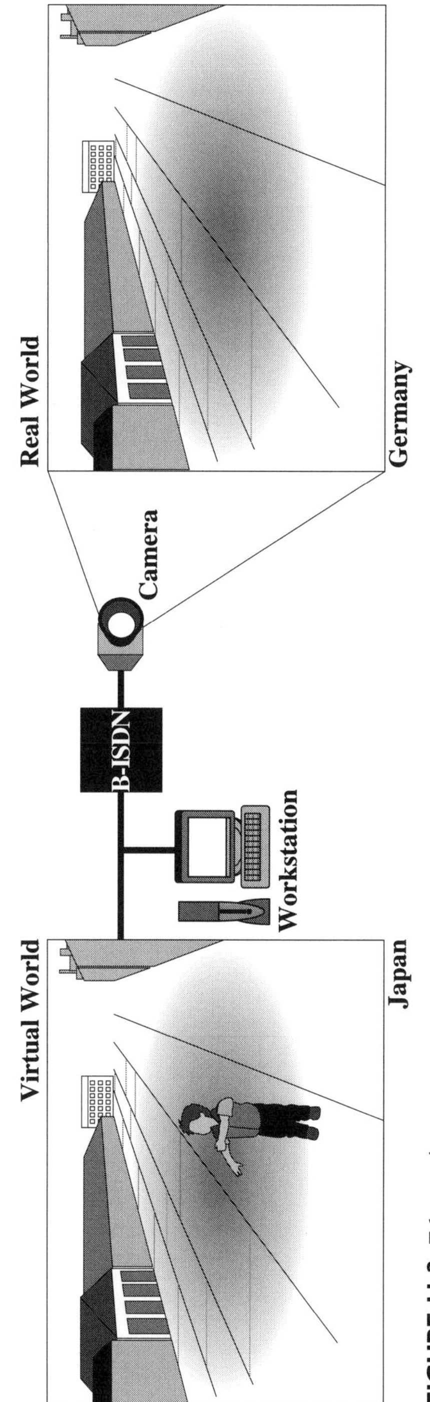

FIGURE 11.2 Telesensation.

FIGURE 11.3 Video murals.

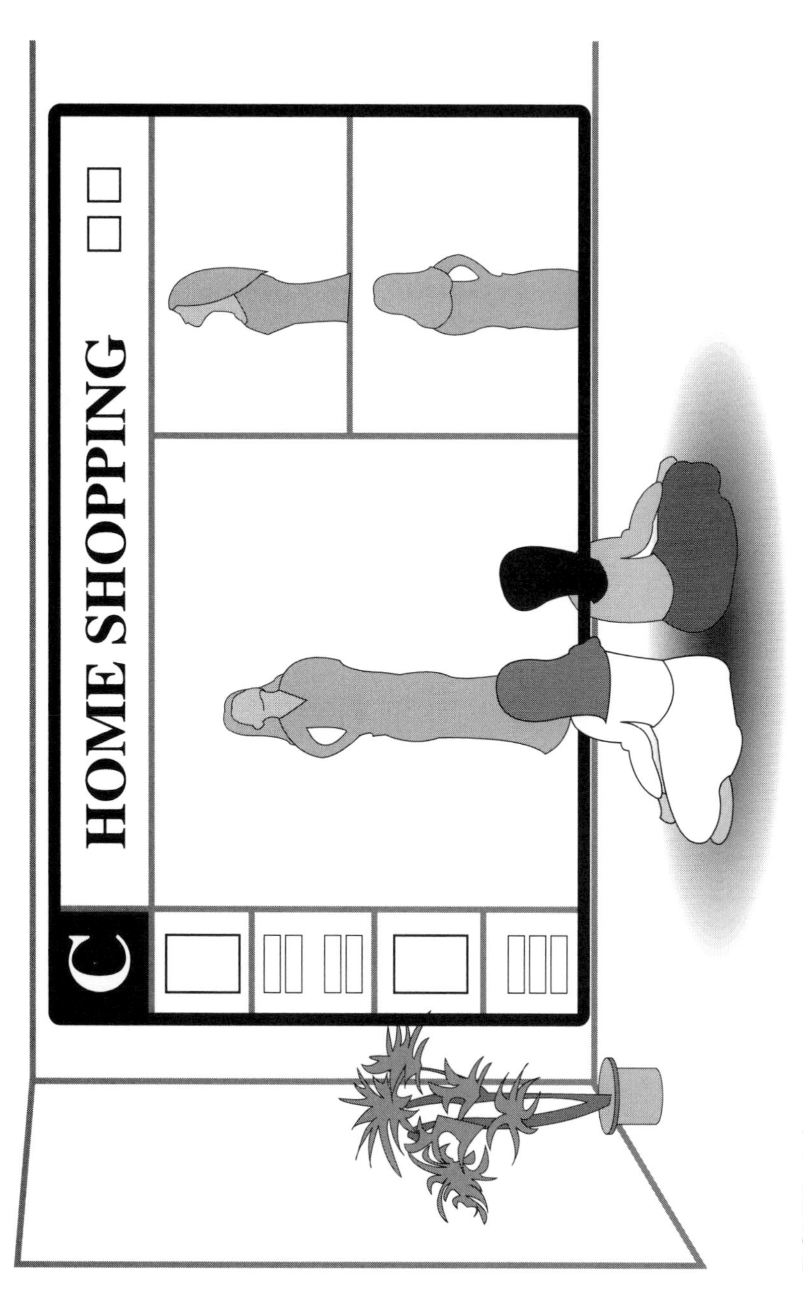

FIGURE 11.4 Teleshopping.

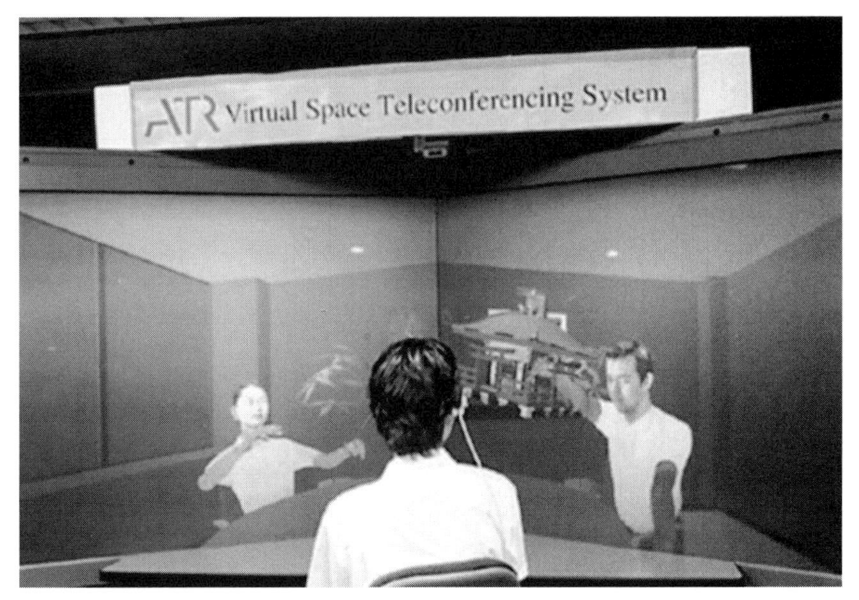

FIGURE 11.5 Virtual space teleconference.

Teleconferencing service: The images of participants who are at different sites are taken by camera, transmitted through communication links to a virtual space, and displayed stereoscopically in the virtual space, where they can have a meeting and carry out cooperative work as if they were gathered in the same place. They can have different views of an object, according to their relative perspectives. They can have a front or side view of the object, according to their viewpoint. They can handle an object by hand gestures and use spoken language, writing, or an image. They can have a stereoscopic view of an object by wearing a special pair of glasses. Figure 11.5 shows a scene of the virtual space teleconferencing system developed by ATR.

11.8 HYPERREALITY

HyperReality (HR), a term coined by Dr. Nobuyoshi Terashima, involves merging VR with actual reality. (*Hyper* means different dimensions.) In HR, the concepts of *HyperWorld* (HW) and coaction fields are introduced. The HW is a seamless world made up of the real and the virtual. *Augmented reality* (AR) is the real world, but augmented. *Extended reality* (ER) is the real world, but extended. Both AR and ER are based on the real world. For example, the structure of a molecule is visualized and displayed stereoscopically on a screen. In the real world we cannot see the structure of a molecule. However, visualization technology enables us to see the structure. This is an example of AR or ER. By comparison, HR is totally different from the real world. For example, in HW we will not be able to distinguish the real world from the virtual world.

One or more coaction fields are introduced in HW. A *coaction field* is a place where inhabitants, such as human beings or animals, real or unreal, work or play together. In a coaction field, the means for communication are provided, including spoken and written language, gesture, or an image. When a human participant discusses something with a computer agent, both the human being and the agent have to share the means for communication. In order for an interaction to take place in a coaction field, there needs to be some common knowledge on the given topic among participants in the same coaction field. For example, in order to talk about traveling, they each need to have common knowledge on the topic of travel. Different attributes can be given to coaction fields, such as physical laws or biological laws. In actual reality, when you throw a ball, the ball falls to the ground according to the law of gravity. However, a coaction field can be made as a nongravity environment if desired. Similarly, in actual reality, a flower grows or wilts with sunlight according to a biological law. Coaction fields can follow their own rules if preferred.

HyperReality, the intermingling of actual reality and VR, comprises the following elements: HyperWorld (HW), inhabitants (I), coaction fields (CF), and visualization (V).

11.8.1 Definitions of HyperReality, HyperWorld, and Inhabitants

HyperReality involves the merging of actual reality with VR. In other words, HR is a seamless world of actual reality and VR. It is composed of HW, I, CF, and V, and can be notated (HW, I, CF, V). The seamless world of HW is made up of the real and the virtual, defined as (RW, VW). It has the following elements:

SCA: Any objects from a natural environment or man-made objects such as a building that are photographed by camera (coded, transmitted through the communication links) and reproduced and displayed stereoscopically by using VR

SCG: Any objects generated by computer graphics (transmitted) and displayed stereoscopically by using VR

SCV: Any objects recognized by computer vision (transmitted, reproduced by computer graphics) and displayed stereoscopically by using VR

A *virtual world* (VW) is notated as (SCA, SCG, SCN). An inhabitant is a real inhabitant (RI) or an artificial inhabitant (AI). A real inhabitant is a human being, an animal, or a plant.

An artificial inhabitant is composed of the following elements:

ICA: An inhabitant, such as a human being, an animal or a plant, photographed by camera (coded, transmitted via communication links, reproduced) and displayed stereoscopically by using VR

ICG: An artificial inhabitant generated by computer graphics (coded, transmitted, reproduced) and displayed stereoscopically by using VR

ICV: An inhabitant, such as a human being, an animal, or a plant, recognized by a computer vision (coded, transmitted, reproduced by computer graphics) and displayed stereoscopically by using VR

AI is notated as (ICA, ICG, ICV).

11.8.2 Coaction Field

One or more coaction fields can be defined in HW. A *coaction field* is a place where inhabitants, such as human beings, animals, or plants, real or artificial, work or play together. To enable interactions, means for communication are provided, such as spoken and written language, gesture, or an image. An object in a coaction field can be under the law of gravity or nongravity. An object can be manipulated by the law of gravity and/or according to a physical law. An object can change its form when it collides with another object. A plant in a coaction field can grow or wilt with sunlight, or such physical laws may be made inapplicable. Coaction fields can be merged or separated dynamically according to interactions among the inhabitants.

11.8.2.1 Definition of Coaction Field

CF = {a field, inhabitants (more than one), means for interaction, law, visualization}

Where field: field in which inhabitants interact
Inhabitants: two and more inhabitants
Means for interaction: spoken and written language, gesture, image, or picture
Physical law: law of gravity and other laws
Biological law, chemical law, and others.
Visualization

11.8.2.2 Creation of Coaction Field

Before a coaction can occur, a coaction field must be created. A coaction field can be created within another coaction field. In the coaction field embedded in another coaction field, the means for interaction, the physical law, or the biological law may be defined. The attributes, such as the means for interaction and the laws in the outer coaction field, can be inherited by the inner coaction field if and only if the attributes of the outer field and those of the inner field do not conflict. If a conflict occurs, the attribute of the inner field is chosen.

11.8.2.3 Integration and Separation of Coaction Fields

The coaction fields CF_i and CF_j can be merged into a new coaction field CF_{ij}, which can be separated again into CF_i, and CF_j. The attributes of Cf_{ij} are inherited from those of Cf_i and Cf_j if and only if the attributes of CF_i and CF_j do not conflict. If a conflict occurs, one of the conflicting rules is chosen and inherited.

11.8.2.4 Examples of Integration of Coaction Fields

In the top most diagram of Figure 11.6, CF_1 has the attributes of knowledge of books and means of reading books. CF_2 has the attributes of the game rules and a knowledge of game machines. In this case, the attributes of both coaction fields do not conflict. The attributes of CF_1 are inherited by CF_2.

In the second diagram of Figure 11.6, CF_1 has the attributes of knowledge of English and the law of gravity. CF_2 has the attributes of knowledge of Japanese and the law of nongravity. When CF_1 and CF_2 merge, the knowledge of English and the knowledge of Japanese are inherited by a new coaction field, CF_3. However, the law of gravity in CF_1 and the law of nongravity in CF_2 come into conflict, so one of these laws needs to be chosen. In the diagram, the law of gravity is selected. In the third diagram of Figure 11.6, the attributes of CF_1 and those of CF_2 do not come into conflict, so all of the attributes in CF_1 and CF_2 are inherited by CF_3.

11.8.2.5 Means for Interaction

The means for interaction among inhabitants are provided in a coaction field.

11.8.2.6 Control Function

The control functions include one that creates, merges, or separates coaction fields and one for interpreting between different languages.

11.8.2.7 Visualization

Although the naked eye cannot normally see a flow of a fluid at the molecular level in the real world, through HR, the flow of a fluid or the structure of a molecule can be visualized and displayed stereoscopically.

11.9 POSSIBLE APPLICATIONS OF HYPERREALITY

The following applications can be developed using HR.

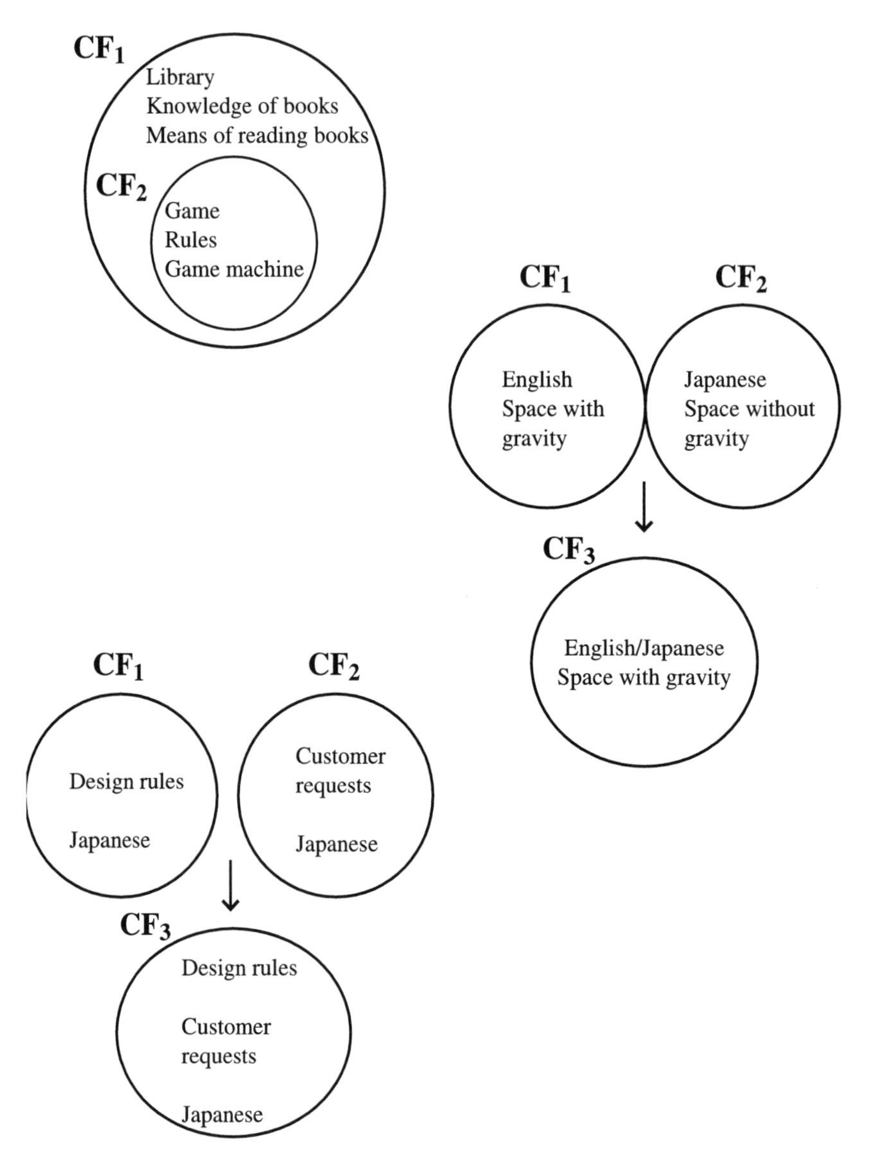

FIGURE 11.6 Diagrams of three coaction fields.

11.9.1 HyperClinic

A doctor and a patient are each located at different sites. Using HR both of them—in reality their images—are brought together in a HyperClinic through communication links. The doctor examines the pulse and temperature of the patient, sees how healthy his face appears, and judges how to treat him. This allows an appropriate medical treatment to be given. The image of a HyperClinic is shown in Figure 11.7.

11.9.2 HyperArt Museum

Each person is at a different location. Through HR, they are brought together in a HyperArt Museum. They can enter, walk through it and enjoy viewing paintings. The image is shown in Figure 11.8.

11.9.3 HyperResort

As shown in Figure 11.9, each person is at a different site, but they are brought together in a HyperResort. They can enjoy sunbathing, play together, and talk to one another.

11.9.4 Manufacturing on Demand

As shown in Figure 11.10, a group of designers is brought together in a Hyper-Design room. They are talking about designing a new car and would like to know users' opinions. HR allows a group of users to be at a HyperMeeting place. They discuss a new car they would like to have. Then both groups—designers and users—are brought together in a common hyperplace where they can interact. Through these interactions, the designers can gather information on users' requirements for a new car. Based on this newly acquired information, the shape or color of the car is changed and shown to the users. When they are satisfied, the new car is manufactured. In this way, manufacturing on demand is realized.

11.9.5 HyperClass

As shown in Figure 11.11, HyperClass is a class where a teacher and students—in reality their images—are brought together through communication links so they can hold classes and carry out cooperative work as if they were gathered in the same place. Anyone from anywhere in the world can access HyperClass and join in a class. The image is shown in Figure 11.11.

In Figure 11.11, two images are shown: one for the teacher and one for the student. In front of them is a virtual object that they are going to handle. Each image's motion is generated by a real human motion. The human is wearing a data glove with sensors that detects any hand motion. In HyperClass, all kinds of virtual objects can be introduced for conducting the class. In the background we see pictures of the students who are attending the class at the remote site. They are watching what the teacher and the student are doing. A diagram of the system is shown in Figure 11.12. In this figure, two sites are shown: the teacher's and the student's. In HyperClass, more than two sites can be interconnected via the Internet. Each site is represented by a person's image.

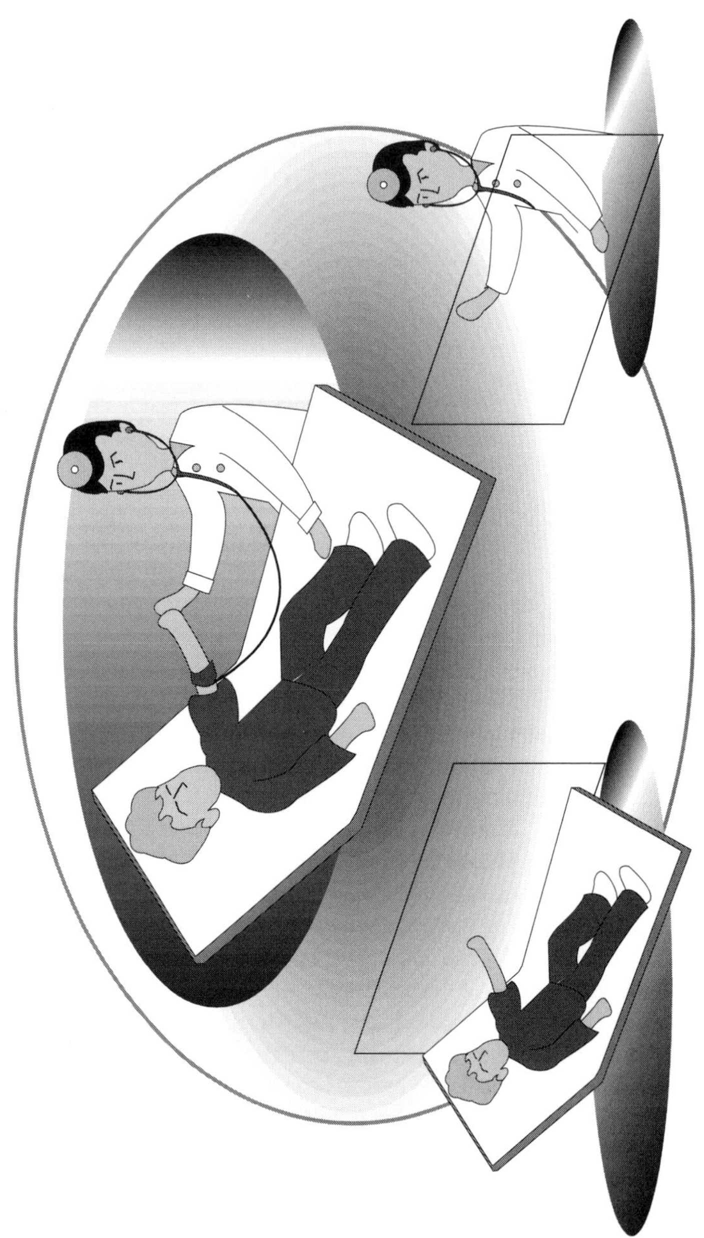

FIGURE 11.7 HyperClinic.

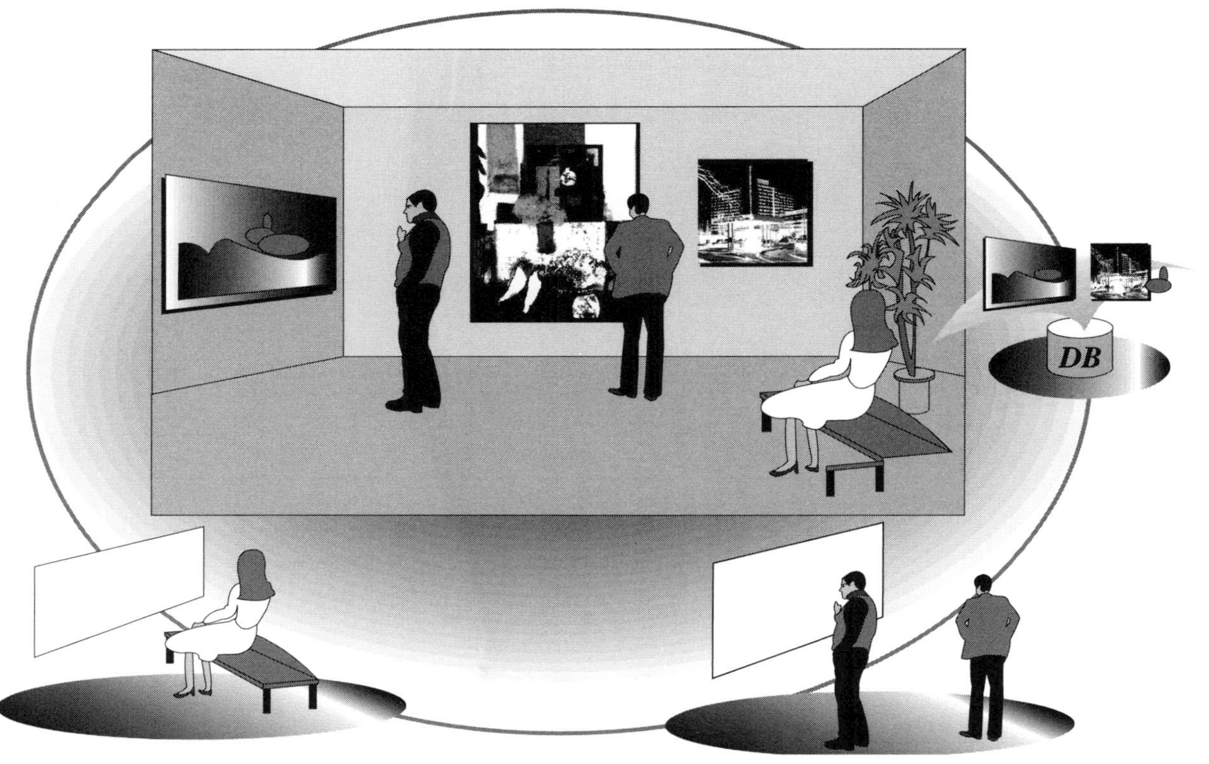

FIGURE 11.8 HyperArt museum.

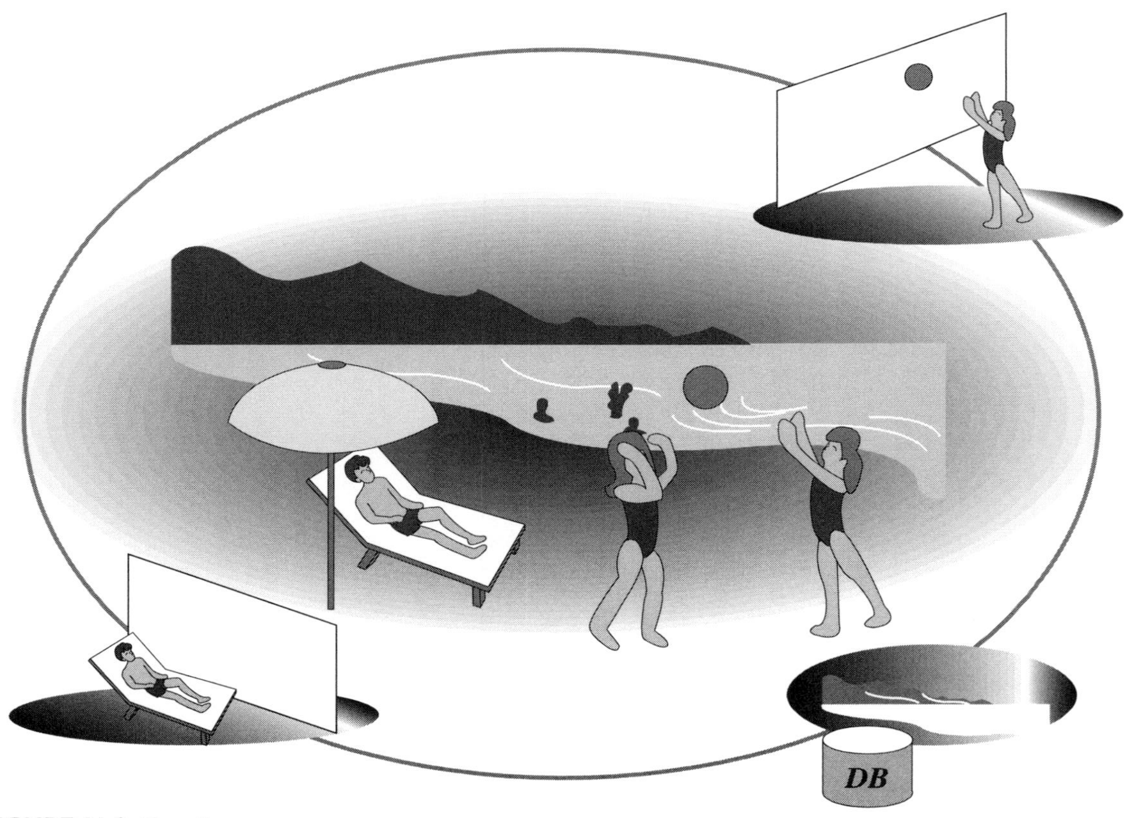

FIGURE 11.9 HyperResort.

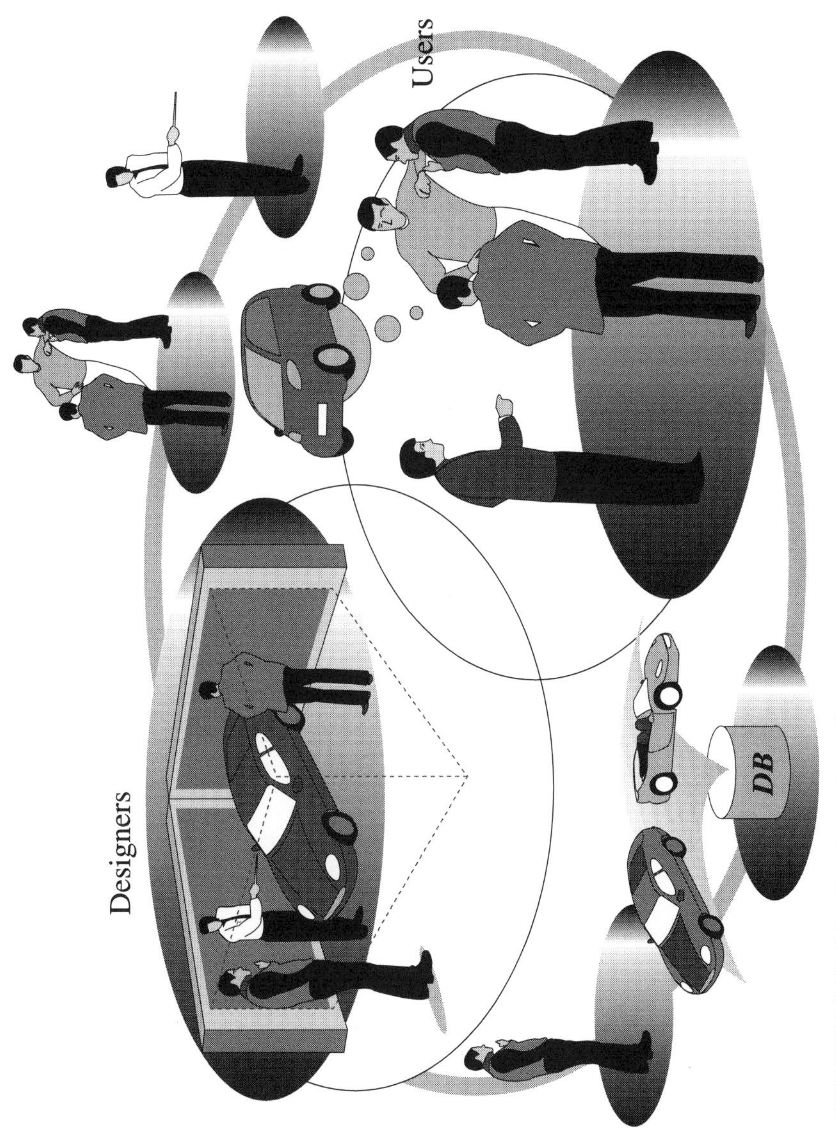

FIGURE 11.10 Manufacturing on demand.

FIGURE 11.11 Scene of HyperClass.

A joint experiment on HyperClass was carried out by connecting Waseda University, Victoria University of Wellington, New Zealand, and the Queensland Open Learning Network, Queensland, Australia over the Internet. One of their tasks was to handle a Japanese artifact. The Japanese teacher talked about Japanese history while handling the virtual object via hand gesture. Then a student of Victoria University and a staff member of the Queensland Open Learning Network handled the object by computer mouse. Another of their tasks was to learn how to assemble a computer from components, by means of hand gesture and the mouse. First, the Japanese teacher showed how to assemble a computer. Then a student and a staff member tried to assemble it. All tasks were conducted in real time.

This system required an intelligent coding technology, which was invented, developed, and installed. The amount of information needed to represent a virtual object such as a Japanese artifact is about 5–10 Mbytes. If the object information is transmitted over the Internet during the class, it is difficult to do this in real-time. To achieve a real-time operation, the intelligent coding technology was installed in HyperClass. The main features of intelligent coding technology are as follows. First the object information, such as a virtual object, teacher's objects, and the class object, are transmitted via the Internet *before* the class starts. During class, only the movement information, such as the object's motion or the teacher's motions, is transmitted. Using this movement information, the virtual object and teachers' objects are adjusted and displayed in HyperClass. Information is transmitted during the class at about 200 bytes/second, so the class was conducted in real time. Wearing shutter glasses gives a stereoscopic view of the objects. In this system a virtual 3D space is created in cyberspace, and anyone can join in and see it together. And they can also handle the virtual object by hand gesture and mouse. This experiment shows that a virtual 3D cyberspace is useful for education.

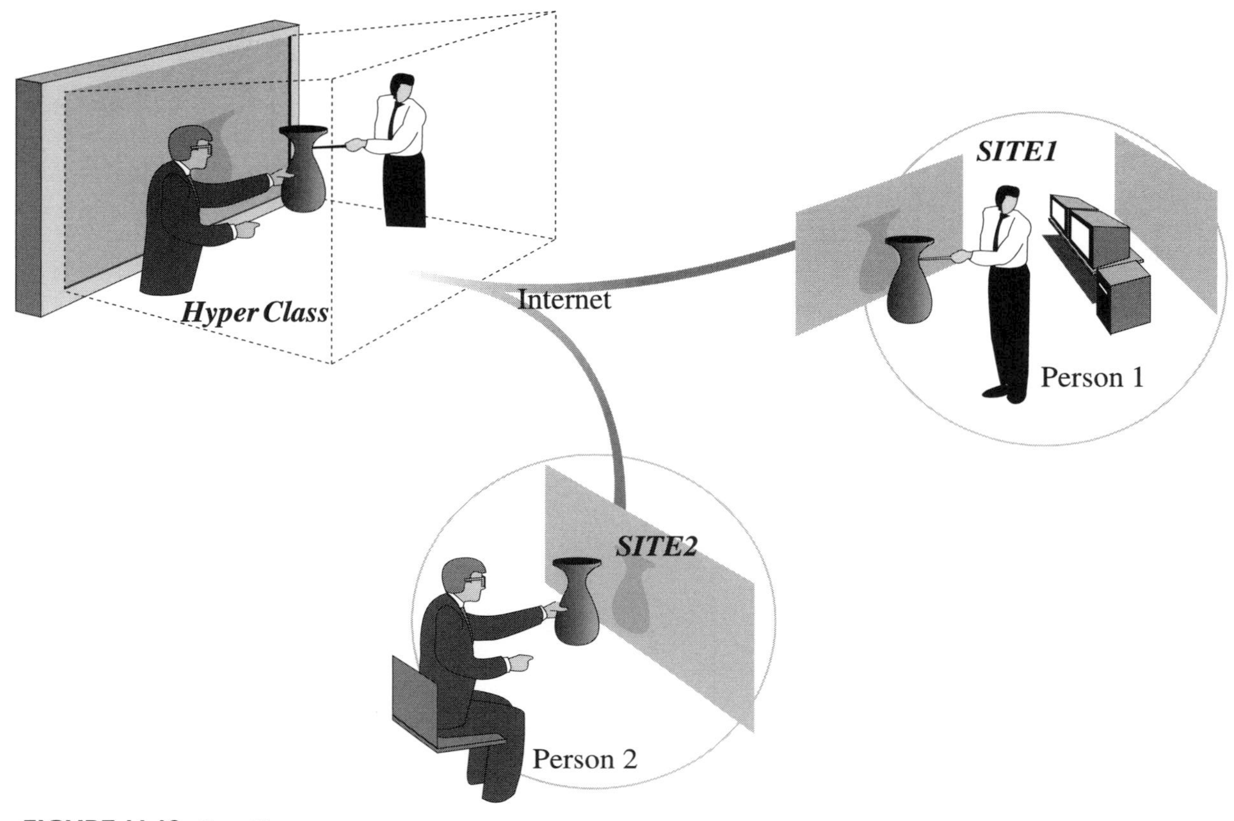

FIGURE 11.12 HyperClass system.

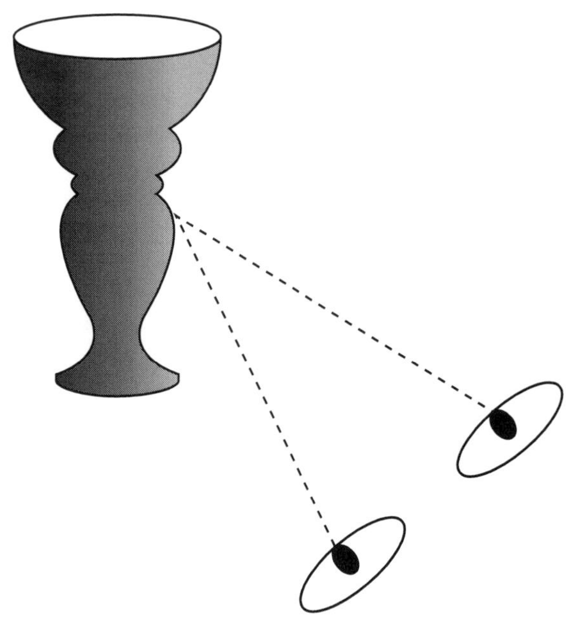

FIGURE 11.13 Binocular vision.

11.10 TECHNOLOGIES FOR ESTABLISHING HYPERREALITY

The main technologies that have to be developed to establish HR are VR, computer vision, and computer graphics.

VR: VR can give us a stereoscopic view of an object when we wear a special pair of glasses. The right view of the object enters the right eye and the left view enters the left eye through the special glasses. The right lens and the left lens are opened one at a time every 8 ms. The right image enters the right eye while the right lens is open and likewise for the left. Viewed with both eyes we see a stereoscopic image, because the right and left views are slightly different. VR enables us to handle an object in a virtual space by means of hand gesture. The shape and texture of the object are stored in a computer. The shape will change according to the hand gestures, and the texture is mapped on the object and displayed in virtual space. Viewers can have a front or a side view of an object, depending on their perspective.

Computer vision: Computer vision is needed to recognize the shape and texture of an object in order to reproduce it. Usually a model of a human face is used to recognize the shape of a human face.

Computer graphics: Using the shape and texture of an object, the object is reproduced to match the viewer's perspective. Using information about hand gestures, the position and shape of the object is adjusted and displayed in virtual space.

12

COMPUTER VISION

Computer vision (CV) is a technology that performs recognition of shape and color of objects that exist in nature such as humans, animals, trees, forests or human-made objects like constructions, handicrafts, cars, or roads. This chapter describes basic theory and shows examples of computer vision.

12.1 DEFINITIONS

12.1.1 Definition of Image Data

Any image can be expressed as two-dimensional data by using the x and y axes. Either the brightness of the image or its gray-level value at the point (x, y) can be expressed by the following equation:

$$f(x, y) = g \quad (1)$$

This equation shows the brightness g of surface-light-emitting devices such as a television screen and a monitor of a personal computer. On the other hand, because images such as photographs and drawings are not self-light-emitting objects, the brightness g of reflective light can be regarded as an image. In this case, g can be expressed by the following equation:

$$g = \log (I_i/I_o) \tag{2}$$

where the amount of incident light is I_i and the amount of reflective light is I_o. So in the case where the image is blackish due to an insufficient amount of reflective light, g is bigger. When the amount of reflective light is too great, making the image whitish, g is a small value.

12.1.2 Definition of a Digital Image

The image given by a continuous gray-level value g on coordinate axes x and y is an *analog* image. The image given by only the gray-level value g at the intersection of coordinate axes x and y is a *sample* image. The image given by a discrete gray-level value of a sample image is a *quantizing* image. The quantizing image is called a *digital* image. In Figure 12.1, assuming $T = 1$ and G is a digital image at coordinate axes (X, Y), the gray-level value $g(X, Y)$ is as given in Eq. (3):

$$g(X, Y) = G \tag{3}$$

$g(X, Y)$ is the minimum-size image and is called a *picture cell.* A digital image is represented as a gray-level value of a picture cell. Therefore the image is represented as the product of M and N, where M is the number of picture cells along the X axis and N is the number of picture cells along the Y axis.

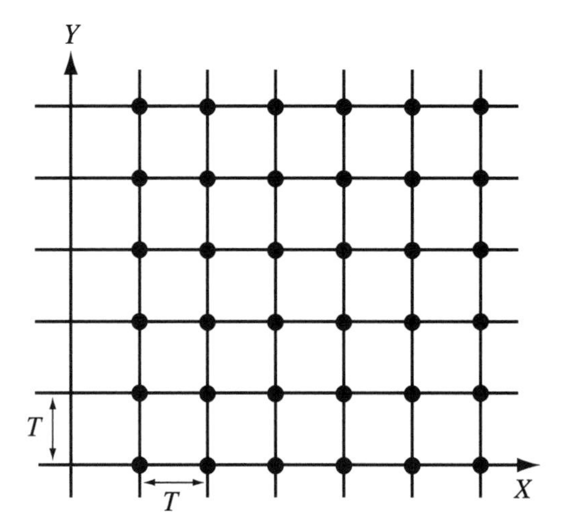

FIGURE 12.1 Sampling.

12.1.3 Quantization of Gray-Level Value

To represent an analog image by discrete gray-level values is called *quantization* of the image. For example, when the brightest gray-level value is "000" and the darkest is "111," this is called a quantization of eight levels. In other words, it is eight gray-scale representations. Conventionally, 64–256 gray-level representations are used for quantization.

12.1.4 Gray-Scale Binary Image Display

If the number of pixels in a region of a gray-scale image is fixed, then the display with binary data, 0 or 1, will be represented by the least amount of data. The features of an image can be preserved by proper conversion to its binary image. When the image is binary data, it can be processed easily.

12.1.5 Threshold Processing

There are a variety of methods for obtaining binary data. One of the easiest uses a gray-level histogram. Consider an image composed of an object and a background. In the density histogram, the gray-level value where there is a large change in the number of pixels is called the *threshold*. (An example is shown in upcoming (Figure 12.4.)

In the second method for obtaining binary data, when the number of pixels of an object is known and the gray level of the object is quite low, the threshold is obtained from the ratio P of the number of pixels of the object to the total number of pixels in the image.

12.2 IMAGE DISPLAY

12.2.1 Density Distribution

The distribution of gray-level values in an image with $M \times N$ pixels is called the *density histogram*. It is not good for image processing if the density distribution is biased. Therefore it is necessary to make the distribution uniform by changing either the light source or the direction of the lens.

12.2.2 Density Transformation

In order to get a good-quality image, a full-scale gray-level display is desirable. To achieve this, the density transformation method is used. As shown in Figure 12.2, density transformation is achieved via line transformation.

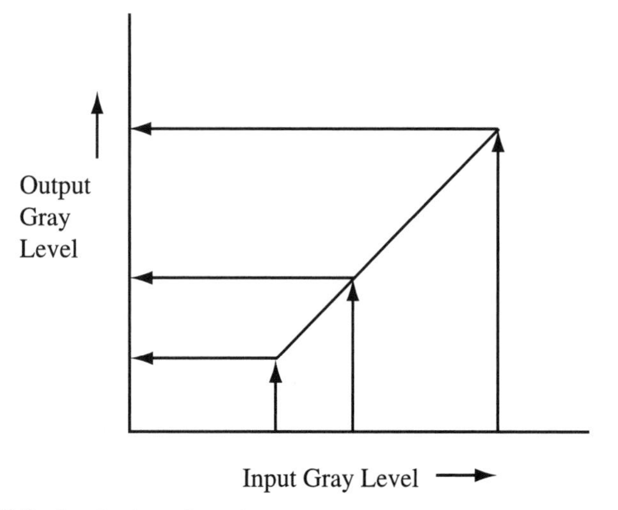

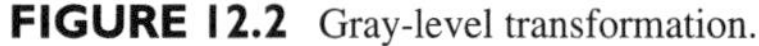

FIGURE 12.2 Gray-level transformation.

12.2.3 Histogram Smoothing

In the case where an object image we would like to recognize is uncertain, histogram smoothing is used for image emphasis (see Figure 12.3). The probability of each gray-level value should be as close as possible. When the number of pixels of the histogram is Q and the number of levels of gray-scale representation is N, then the average number of pixels for each level is Q/N. Summation of pixels is performed until the number of pixels reaches the average number. In this case, the number of pixels nearest to the average number is chosen whether the former is bigger or smaller than the latter. This operation continues until the maximum gray-level value is reached.

New gray level values are $G_{MIN} + D/N$, $G_{MIN} + 2 \times D/N, \ldots, G_{MAX}$ where N is the number of levels of gray-level values and D is the difference between the maximum gray-level value, G_{MAX}, and the minimum gray-level value, G_{MIN}.

12.2.4 Gray-Scale Image Display

There is a limitation when expressing a gray-scale level on a CRT display or other type of screen. Using the binary gray-scale level image display, methods for displaying an image that is similar to the original one have been developed. A typical one is the dither method.

12.2.5 Binary Dither Method

In this method, by comparing a threshold value $B(x, y)$ with an input image gray level $g(x, y)$, we obtain a binary value 1 or 0. The dither method is shown in

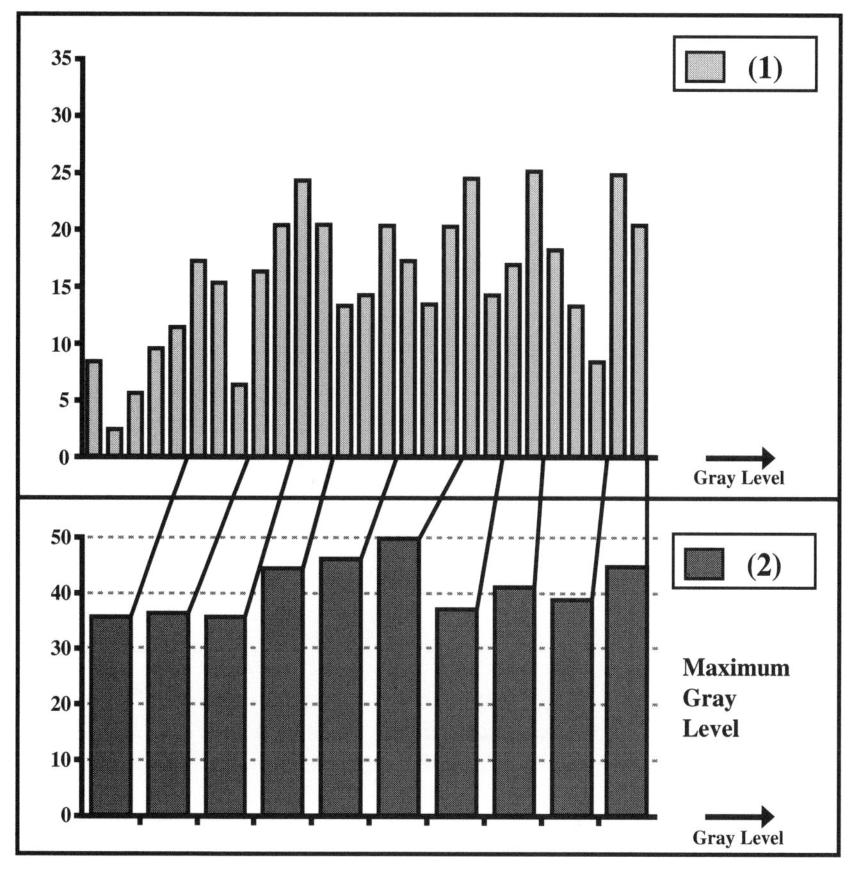

FIGURE 12.3 Smoothing of a histogram.

Figure 12.5. A binary value is obtained as follows:

$$GB(x, y) = \begin{cases} 1 & \text{if } g(x, y) \geq B(x, y) \\ 0 & \text{if } g(x, y) < B(x, y) \end{cases} \tag{4}$$

12.2.6 Organizational Dither Method

The organizational dither method decides the binary value of a pixel by considering the coordinate information of the pixel. We compare the value of the pixel with the corresponding value of a dither matrix. The value 1 is assigned to the pixel when its value is greater than or equal to the corresponding value of the dither matrix; the value 0 is assigned when the pixel's value is smaller than the corresponding value of a dither matrix, where the value $g(i, j)$ of a pixel is normalized to $0 \leq g(i, j) \leq 15$.

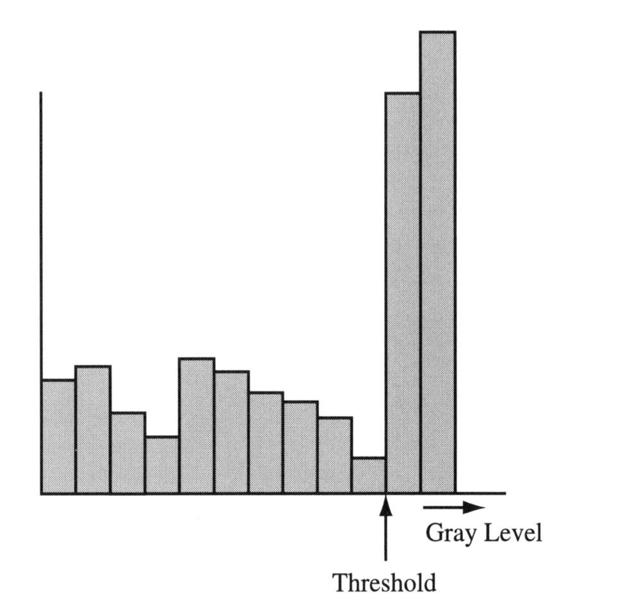

FIGURE 12.4 Threshold.

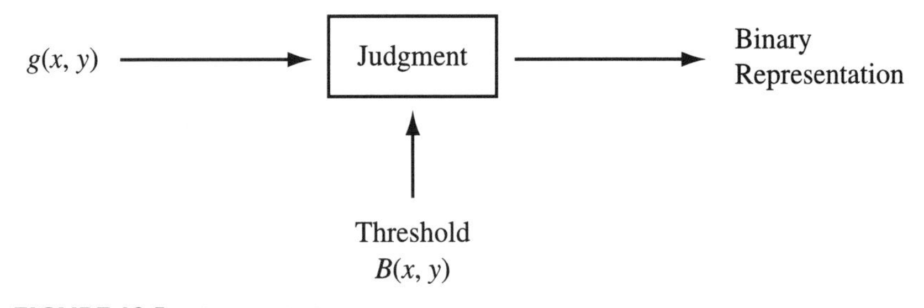

FIGURE 12.5 Dither Method.

Given the element of a dither matrix $T(i, j)$, the value of pixel $g_r(i, j)$ is determined as follows.

$$g_r(i, j) = \begin{cases} 1 & (g(i, j) \geq T(i, j)) \\ 0 & (g(i, j) < T(i, j)) \end{cases} \tag{5}$$

An example is shown in Figure 12.6.

12.2.7 Multivalue Dither Method

In addition to the binary dither method, the three-value and four-value dither methods have been proposed. These are extensions of the binary dither method. In the three-value dither method, a value of a half-level is taken into consideration.

2	10	0	12
10	5	15	8
4	12	3	10
10	10	15	4

(a) Input Image

0	8	2	10
12	4	14	6
3	11	1	9
15	7	13	5

(b) Dither Matrix

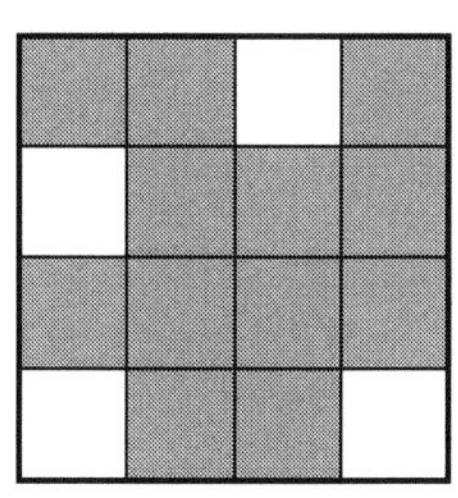

(c) Output Image

FIGURE 12.6 Dither matrix.

The binary dither method is applied to pixels with a gray-level value of 0 to a half, and 0 or a half is assigned to the value of the pixel.

In the same way, the method is applied to pixels with a gray-level value of a half to 1, and a half or 1 is assigned to the value of a pixel. The four-value dither method is applied to the pixel with a value of 0 to one-third, one-third to two-thirds, and two-thirds to 1.

12.2.8 Color Compression

We shall present one of the easiest color compression algorithms. Consider an image with 8-bit gray level for r, g, and b. The region of color, which exists in RGB space created along orthogonal axes r, g, and b, is divided into 256 regions as follows. An image is processed four times. From the first time to the third, the region is divided. The fourth time, a color is assigned to the pixel.

The compression process steps are performed in the following sequence.

(1) The maximum value and the minimum value of r are determined, and the difference between the two values is divided into eight regions.
(2) The difference between the maximum and minimum values of g is divided into eight regions. Then the region of r and g is divided into 64 regions.
(3) The difference between the maximum and minimum values of b is divided into eight regions. Then the region of r, g, and b is divided into 256 regions.
(4) A decision is made as to which region of r, g, and b each pixel belongs to, and the pixel is represented by the (r, g, b) value of the selected region.
(5) Each region is represented by the mean value of r, g, and b in the region.

12.3 IMAGE TRANSFORMATION

Mathematical methods for two-dimensional image transformation are described in this section. By one method, digital image transformation is performed via a space filtering operation. In another method, the filtering operation is performed by conversion from the space domain to the spectrum domain.

$(i-1,\ j-1)$	$(i,\ j-1)$	$(i+1,\ j-1)$
$(i-1,\ j)$	$(i,\ j)$	$(i+1,\ j)$
$(i-1,\ j+1)$	$(i,\ j+1)$	$(i+1,\ j+1)$

FIGURE 12.7 3×3 region for smoothing.

12.3.1 Space Filtering

This method has the following objectives: (1) to eliminate noise from the source image, called *smoothing*, and (2) to detect features of the image such as contour and direction from the source image. To achieve these objectives we use the Laplacian and gradient operations.

When it is difficult to obtain gray-level values accurately because a high frequency has been introduced into the video signal, a smoothing operation is performed. A mean value is calculated for nine pixels, from pixel $(i-1,\ j-1)$ to pixel $(i+1,\ j+1)$, and this value is assigned to a new gray level (see Figure 12.7). The formula is as follows:

$$T(i,j) = \frac{1}{9}\begin{bmatrix} 1 & 1 & 1 \\ 1 & 1 & 1 \\ 1 & 1 & 1 \end{bmatrix} \tag{6}$$

$$f(i,j) = g(i,j) \times T(i,j)$$

where $T(i,j)$ is the coefficient of $g(i,j)$. This means that $f(i,j)$ is the summation from $g(i-1,j-1)$ to $g(i+1,j+1)$ divided by 9. The gray-level values $g(i-1,j-1)$, $g(i,j-1)$, $g(i+1,j-1)$, $g(i-1,j)$, $g(i+1,j)$, $g(i-1,j+1)$, $g(i,j+1)$, $g(i+1,j+1)$ are used to calculate $f(i,j)$ in space filtering.

In space filtering, a mean value is used for $f(i,j)$. Another method involves decreasing the value of the coefficient as the distance from $g(i,j)$ increases. One example is shown in Eq. (7).

$$T(i,j) = \frac{1}{10}\begin{bmatrix} 1 & 1 & 1 \\ 1 & 2 & 1 \\ 1 & 1 & 1 \end{bmatrix} \tag{7}$$

In general, $T(i,j)$ is described as follows:

$$T(i,j) = \begin{bmatrix} w3 & w2 & w1 \\ w4 & w0 & w8 \\ w5 & w6 & w7 \end{bmatrix} \tag{8}$$

Here, w_n is the coefficient of the gray level.

12.3.2 Laplacian Contour Detection

Whereas the main objective of space filtering is to eliminate noise through an integration operation, contour detection is performed through a differential operation on the image. One of the procedures is as follows.

(1) The square of the difference between $g(i+1, j)$ and $g(i, j)$ and the square of the difference between $g(i, j)$ and $g(i, j+1)$ are added together.
(2) The square root of the value obtained in step (1) is assigned to $f(i, j)$. The operation is shown in Eq. (9):

$$f(i,j)=[\{g(i+1,j)-g(i,j)\}^2+\{g(i,j)-g(i,j+1)\}^2]^{1/2} \tag{9}$$

where $g(i, j)$ is the gray-level value of pixel (i, j).

One drawback of this differential operation is that it emphasizes noise included in the image. To overcome this, a Laplacian operation is introduced. This corresponds to the secondary differential operation: $f(i, j)$ is obtained via the term

$$\{-g(i, j-1) - g(i-1, j) + 4g(i, j) - g(i+1, j) - g(i, j+1)\}.$$

The Laplacian coefficient is given in Eq. (10)

$$\pounds(i,j)=\begin{bmatrix} 0 & -1 & 0 \\ -1 & 4 & -1 \\ 0 & -1 & 0 \end{bmatrix} \tag{10}$$

By another method, $f(i, j)$ is obtained using the term

$$\{-g(i-1, j-1) - g(i-1, j) - g(i-1, j+1) - g(i-1, j) + 8g(i, j) \\ - g(i+1, j) - g(i+1, j-1) - g(i+1, j) - g(i+1, j+1)\}$$

The Laplacian coefficient is given in Eq. (11).

$$\pounds(i,j)=\begin{bmatrix} -1 & -1 & -1 \\ -1 & 8 & -1 \\ -1 & -1 & -1 \end{bmatrix} \tag{11}$$

In both of these methods, the contour is determined. However, the direction of contour cannot be obtained. To determine the direction of contour, methods such as the Sobel operator are used.

12.3.3 Filters for Special Purposes

There are filters that detect specific shapes. The operation is the same as with template matching.

12.3.3.1 Line Detection

Using the following filter, a line is detected.

$$
\begin{aligned}
Vertical\ Line &= \begin{bmatrix} -1 & 1 & -1 \\ -1 & 1 & -1 \\ -1 & 1 & -1 \end{bmatrix} \\
Horizontal\ Line &= \begin{bmatrix} -1 & -1 & -1 \\ 1 & 1 & 1 \\ -1 & -1 & -1 \end{bmatrix}
\end{aligned}
\tag{12}
$$

12.3.3.2 Determining Contour and Direction

Contour and direction of an image are determined using a Sobel operator. Sobel operators are given by Δx_n and Δy_n. The parameter n shows the domain of the filtering operation. At $n = 1$, the filtering operation is performed for a 3×3 domain. At $n = 2$, an operation is performed for a 5×5 domain. At $n = n$, the filtering operation is performed for a $(2n + 1) \times (2n + 1)$ domain. For example, Sobel operators Δx_1 and Δy_1 are given in Eqs. (13) and (14), respectively:

$$
\Delta x_1 = \begin{bmatrix} -1 & 0 & 1 \\ -2 & 0 & 2 \\ -1 & 0 & 1 \end{bmatrix} \tag{13}
$$

$$
\Delta y_1 = \begin{bmatrix} -1 & -2 & -1 \\ 0 & 0 & 0 \\ 1 & 2 & 1 \end{bmatrix} \tag{14}
$$

When an input image signal $g(i, j)$ $(0 \le i, j \le N - 1)$ is given, output signals $\Delta X_1(i, j)$ and $\Delta Y_1(i, j)$ are calculated by using filtering operators Δx_1 and Δy_1, respectively, where N is input image size. $\Delta x_1(i, j)$ and $\Delta y_1(i, j)$ are given in Eqs. (15) and (16), respectively:

$$
\Delta X_1(i,j) = \sum_{v=-1}^{1} \sum_{w=-1}^{1} \Delta x_1(v,w) \cdot g(v+i,\ w+j) \tag{15}
$$

$$
\Delta Y_1(i,j) = \sum_{v=-1}^{1} \sum_{w=-1}^{1} \Delta y_1(v,w) \cdot g(v+i, w+j) \tag{16}
$$

where $1 \le i, j \le N - 2$.

Finally, an output image signal $G(i, j)$ is given in Eq. (17).

$$G(i,j) = \{\Delta X_1(i,j)^2 + \Delta Y_1(i,j)^2\}^{1/2} \tag{17}$$

Direction is given in Eq. (18).

$$\Theta(i,j) = \tan^{-1}\{\Delta X_1(i,j)/\Delta Y_1(i,j)\} \tag{18}$$

Now we present an example.

An input signal is as follows:

$$\begin{bmatrix} 1 & 1 & 2 \\ 1 & 2 & 2 \\ 2 & 2 & 3 \end{bmatrix}$$

$\Delta x_1(i, j)$, $\Delta y_1(i, j)$, $G(i, j)$, and $\theta(i, j)$ are as follows:

$$\Delta y_1(i,j) = 4$$

$$G(i,j) = 4\sqrt{2}$$

$$\theta(i,j) = \tan^{-1}(4/4) = 45°$$

In general Δx_n and Δy_n are as follows:

$$\Delta X_n(i,j) = \sum_{v=-n}^{n}\sum_{w=-n}^{n} \Delta x_n(v,w)\cdot g(v+i,w+j) \tag{19}$$

$$\Delta Y_n(i,j) = \sum_{v=-n}^{n}\sum_{w=-n}^{n} \Delta y_n(v,w)\cdot g(v+i,w+j) \tag{20}$$

where $n \le i, j \le N - n - 1$.

Finally, an output image signal $G(i, j)$ is as given in Eq. (21).

$$G(i,j) = \{\Delta X_n(i,j)^2 + \Delta Y_n(i,j)^2\}^{1/2} \tag{21}$$

The direction angle $\theta(i, j)$ is given in Eq. (22).

$$\theta(i,j) = \tan^{-1}\{\Delta X_n(i,j)/\Delta Y_n(i,j)\} \tag{22}$$

12.3.4 Spectrum Transform

The transformation from the space domain to the frequency domain for an input signal is performed, and the frequency spectrum is determined. Then a filtering operation is performed on the spectrum.

To obtain the output image, an inverse frequency transform for the spectrum is performed.

12.3.4.1 Fourier Transform

If function $g(t)$ is a continuous function with period T, then $g(t)$ can be represented as direct current and components with frequency n/T ($n = 1, 2,\ldots$)

$$\begin{aligned} g(t) &= a_0 + a_1 \cos 2\pi t/T + a_2 \cos 4\pi t/T + a_3 \cos 6\pi t/T + \cdots \\ &\quad + b_1 \sin 2\pi t/T + b_2 \sin 4\pi t/T + b_3 \sin 6\pi t/T + \cdots \\ &= a_0 + \sum_{n=1}^{\infty} a_n \cos 2\pi nt/T + \sum_{n=1}^{\infty} b_n \sin 2\pi nt/T \end{aligned} \tag{23}$$

where, a_0, a_n, b_n are Fourier coefficients.

Fourier coefficients a_0, a_n, b_n are given in Eqs. (24), (25), and (26), respectively.

$$a_0 = \frac{1}{T}\int_0^T g(t)dt \tag{24}$$

$$a_n = \frac{2}{T}\int_0^T g(t)(\cos 2\pi nt/T)dt \tag{25}$$

$$b_n = \frac{2}{T}\int_0^T g(t)(\sin 2\pi nt/T)dt \tag{26}$$

12.3.4.2 Discrete Fourier Transform

When the period of a signal is $T = N$, the base frequency is $f = 1/N$, and the high frequency is $k/N(k = 0, 1, 2,\ldots, N-1)$, the Fourier expressions $a(1/N)$ and $b(1/N)$ are as given in Eqs. (27) and (28), respectively.

$$a(1/N) = \frac{1}{N}\sum_{k=0}^{N-1} g(k)(\cos -2\pi k/N) \tag{27}$$

$$b(1/N) = \frac{1}{N}\sum_{k=0}^{N-1} g(k)(\sin -2\pi k/N) \tag{28}$$

In general, $a(n/N)$ and $b(n/N)$ are as given in Eqs. (29) and (30), respectively.

$$a(n/N) = \frac{1}{N}\sum_{k=0}^{N-1} g(k)(\cos - 2\pi kn/N) \quad (29)$$

$$b(n/N) = \frac{1}{N}\sum_{k=0}^{N-1} g(k)(\sin - 2\pi kn/N) \quad (30)$$

Direct current $a(0)$ is as given in Eq. (31).

$$a(0) = \frac{1}{N}\sum_{k=0}^{N-1} g(k) \quad (31)$$

The inverse Fourier transform is given in Eq. (32).

$$g(n) = a(0) + \sum_{k=1}^{N-1} a(k/N)\cos(2\pi kn/N) + \sum_{k=1}^{N-1} b(k/N)\sin(2\pi kn/N) \quad (32)$$

The Fourier transform is expressed using complex numbers as follows:

$$G(n/N) = \frac{1}{N}\sum_{k=0}^{N-1} g(k)\exp(-j2\pi nk/N) \quad (33)$$

The inverse Fourier equation is as follows:

$$g(n) = \sum_{k=0}^{N-1} G(k/N)\exp(j2\pi nk/N) \quad (34)$$

Here a rotator ϕ is introduced, as expressed in Eq. (35).

$$\phi = \exp(-j2\pi/N) \quad (35)$$

$$\exp(a \times b) = (\exp a)b \quad (36)$$

The following equations are introduced by using Eqs. (35) and (36).

$$\begin{aligned} \exp(-j2\pi nk/N) &= \Phi^{nk} \\ \exp(j2\pi nk/N) &= \Phi^{-nk} \end{aligned} \quad (37)$$

Using Eq. (37), Eqs. (33) and (34) become Eqs. (38) and (39), respectively.

$$G(n/N) = \frac{1}{N}\sum_{k=0}^{N-1} g(k)\Phi^{nk} \tag{38}$$

$$g(n) = \sum_{k=0}^{N-1} G(k/N)\Phi^{-nk} \tag{39}$$

In general, $G(n'_1, n'_2, \ldots, n'_m)$ is shown in Eq. (40).

$$G(n'_1, n'_2, \ldots, n'_m) = \sum_{n_1=0}^{N_1-1}\sum_{n_2=0}^{N_2-1}\cdots\sum_{n_m=0}^{N_m-1} g(n_1,\ n_2, \ldots, n_m)\Phi_{N_1}^{n_1 n'_1}\Phi_{N_2}^{n_2 n'_2}\cdots\Phi_{N_m}^{n_m n'_m} \tag{40}$$

where

$$\Phi_{N_i}^{n_m n'_m} = \exp(-j2\pi n_m n'_m / N_i) \tag{41}$$

12.3.4.3 Spectrum Analysis by Means of an Orthogonal Function

The Walsh function has a value of 1 or −1. It is used for spectrum analysis of discrete data. The Walsh function is composed of wal(m, n), where $m = 0, 1, \ldots, N-1$ and $n = 0, 1, \ldots, N-1$. The Walsh function with $N = 8$ is shown in Figure 12.8. In general, the Walsh function is as follows:

$$\text{wal}(0, n) = 1 \qquad (n = 0, 1, 2, \ldots, N-1)$$

$$\text{wal}(1, n) = \begin{Bmatrix} 1(n = 0,\ 1,\ 2, \ldots,\ N/2-1) \\ -1(n = (N/2),\ (N/2)\ +\ 2, \ldots,\ N-1) \end{Bmatrix} \tag{42}$$

$$\text{wal}(m, n) = \text{wal}([m/2], 2n)\ \text{wal}(m - 2[m/2], n)$$

where [☒] is Gaussian and is an integer value of [m/2].

In Figure 12.8,

wal(0, n) is a direct current
wal(1, n) wal(3, n) are the base wave
wal(2, n) wal(4, n) are the twofold wave
wal(7, n) wal(0, n) are the quadruple wave

where m is frequency, which varies between 1 and −1. $m = 1$ means that frequency is once and $m = 2$ means that frequency is twice.

The two-dimensional Walsh function wal(l, k) is given in Eq. (43).

$$W_{ml}(n, k) = \text{wal}(m, n) \times \text{wal}(1, k) \tag{43}$$

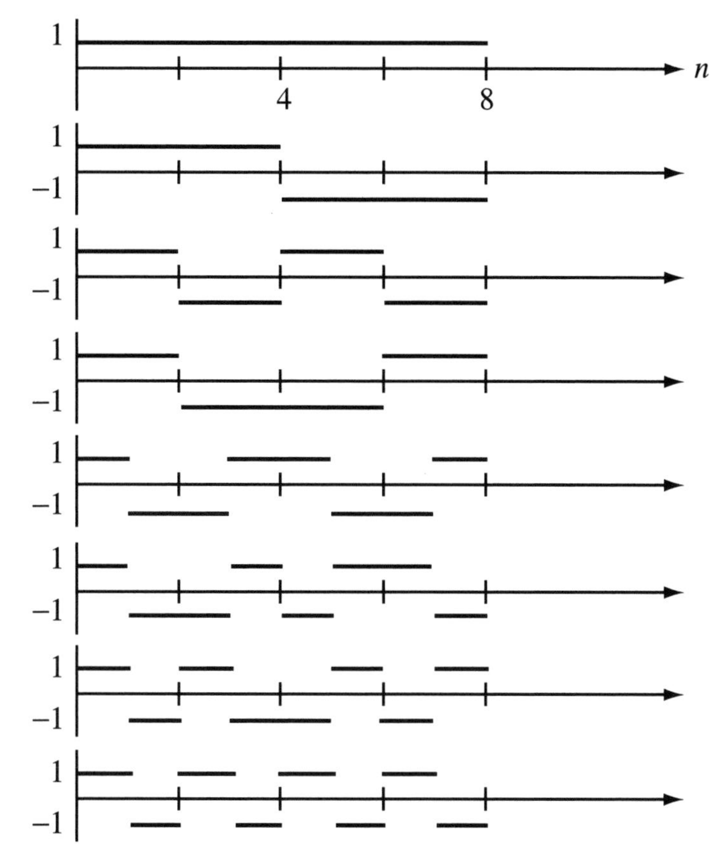

FIGURE 12.8 Wave form of Walsh function.

Figure 12.9 shows an example of $W_{ml}(n, k)$. When $N = K = 4$, 16 functions from w_{00} to w_{33} of $W_{ml}(n, k)$ are obtained.

The Walsh transform spectrum for image $g(n, k)$ is shown in Eq. (44).

$$G(m,l) = \sum_{n=0}^{N-1} \sum_{k=0}^{K-1} g(n,k) W_{ml}(n,k) \tag{44}$$

$G(m, l)$ is a two-dimensional Walsh spectrum and is used for the feature parameter of a two-dimensional image $g(n, k)$. The coefficients of the Walsh function show the feature of an image and are used for image analysis. When image analysis is performed, the Walsh coefficients of an image are calculated and compared with those of a sample image, and then the image is analyzed.

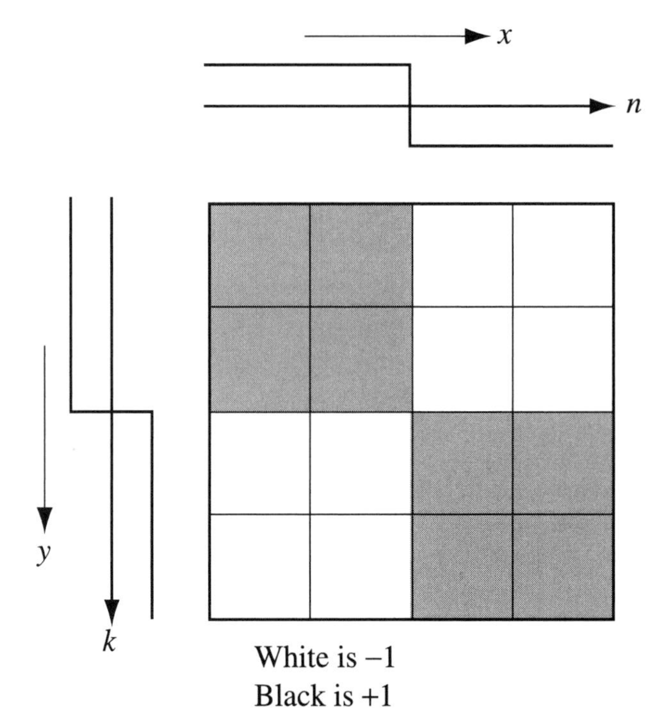

FIGURE 12.9 Example of $W_{ml}(n, k)$.

12.4 IMAGE RECOGNITION FOR TELESENSATION

12.4.1 Image Analysis

12.4.1.1 Line Analysis

The important information that is included in a line segment helps to analyze a figure or an image. To recognize a rectangular solid object, line segments of the object need to be determined. Using the structural information about line segments, the shape of the object is recognized. For a binary image, the value of any pixel brighter than the specific gray level is set at 0, and the value of a pixel darker than the specific value is set at 1.

12.4.1.2 Assignment of Multiple Threshold Values

For a binary image, the method employs not a single threshold value, but multiple threshold values. A binary image generated by multiple threshold values is considered a level plane. Using level planes, all of the contour lines are clear.

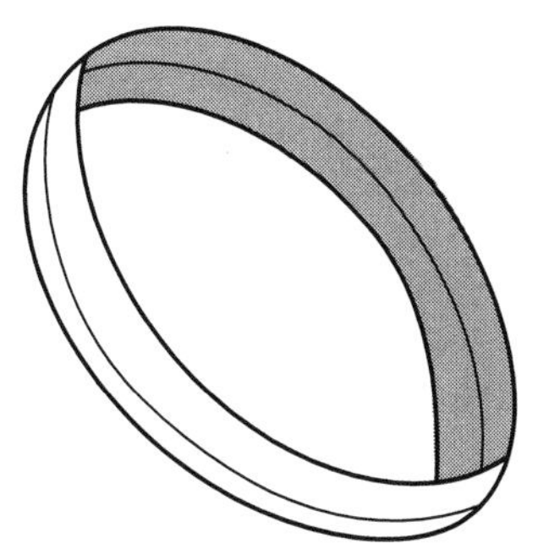

FIGURE 12.10 Centerline extraction.

12.4.1.3 Line Thinning

Line thinning is used to establish the characteristics of a binary image. In more concrete terms, a contour line with a width of 1 pixel is obtained from a contour line with a larger width by progressively eliminating the contour line's pixels, from the outside in. Figure 12.10 shows the image of centerline extraction.

12.4.1.4 Digital Image

With a digital image, a point has a domain of 1 pixel and a line is a connection of such points, in which case, the points are called *connected.* Digital images have either 4-connected images or 8-connected images.

4-neighbors and 8-neighbors: Figure 12.11 shows eight pixels surrounding point P_0. These pixels are called the *8-neighbors*. The four pixels just above P_0, just below P_0, just left of P_0, and just right of P_0 are the called *4-neighbors*. An image is connected upward if P_2 is 1 and connected downward if P_6 is 1. When pixels from P_1 to P_8 are 0, P_1 is called an *isolation point*. When some of the pixels from P_1 to P_7 are 1, P_0 is a *boundary point*. When all pixels from P_1 to P_8 are 1, then P_0 is an *internal point*.

4-connected and 8-connected: When a pixel M is connected to a pixel N with 4-neighbors, M and N are said to be *4-connected*. Figure 12.12 presents an example of being 4-connected from M to N, where a_1, a_2, a_3, and a_4 are the four paths. All of the pixels that occupy the four paths are called the components of the 4-connected. When M and P are 8-connected, M is connected to P through pixels with 8-neighbors. In Figure 12.12, M and P are *8-connected*, where a_1, a_2, a_3, a_4, and a_5 are the eight paths.

Connectivity: The number of connected components that are separated from each other is called the *connectivity*. In Figure 12.12, the connectivity is 1 in the case of the 8-connected and the connectivity is 3 in the case of the 4-connected.

P_3	P_2	P_1
P_4	P_0	P_8
P_5	P_6	P_7

FIGURE 12.11 4-neighbors and 8-neighbors.

			M			
		a_2	a_1			
		a_3				
	N	a_4				
			a_5			
				P		

FIGURE 12.12 Chain of pixels.

12.4.1.5 Texture Analysis

A texture is an iterative pattern put in order, for example, a pattern of a piece of cloth is a texture. Texture analysis is performed as follows.

(1) Texture is recognized by means of texture analysis, which detects an ordered image pattern or chooses a different domain from a pattern. For instance, a desert or a house is differentiated from a field via texture analysis.
(2) When a difference among textures is detected, a border between the two domains is chosen.
(3) A texture is mapped on an image, and image reproduction is performed.

12.4.1.6 Detection of Texture

As shown in Figure 12.13, an area can be subdivided into smaller areas. This is accomplished as follows. A mean value and a dispersion value of the gray level of a small area are calculated. Given a mean value M_i and a dispersion value of V_i

$T(1, 1)$	$T(2, 1)$	...	$T(N, 1)$
$T(1, 2)$	$T(2, 2)$	...	$T(N, 2)$
		...	
		⋮ ⋮	
$T(1, N)$	$T(2, N)$		$T(N, N)$

FIGURE 12.13 Area segmentation.

for each small area, a homogeneous larger area is created by integrating the small areas that satisfy the following:

$$p_1 < M_i < p_2 \qquad \text{and} \qquad q_1 < V_i < q_2$$

Conventionally, 3×3 or 5×5 small areas are used for integration.

Another method creates a homogeneous larger area by using a power spectrum of a small area. Given the power spectrum G_{ij} of a region with $N \times N$ pixels, G is as defined in Eq. 45:

$$G = (G_{00}, G_{01}, G_{02}, \ldots, G_{kk}) \qquad k = N/2 - 1 \tag{45}$$

where G is a feature vector and shows a texture in the area (00–kk). Given a feature vector G in each area, the distance between areas k and l is as defined in Eq. (46):

$$d_{kl} = | G_k - G_l |^2 \tag{46}$$

where G_k, G_l are feature vectors of areas k and l, respectively.

12.4.2 Image Compression

Image compression is useful for image analysis, image storage, and image transmission. The compression is accomplished by using a pyramid data structure or a tree structure.

12.4.2.1 Method Using a Pyramid Data Structure

A series of images whose original image size is $1/(2^{2n})$ ($n = 1, 2, \ldots$) is called a pyramid data structure. As shown in Figure 12.14, in the case of a binary image, the value of a pixel on a higher level is determined by using the conjunction of the values of the four pixels on the lower level. For example, if at least one pixel with

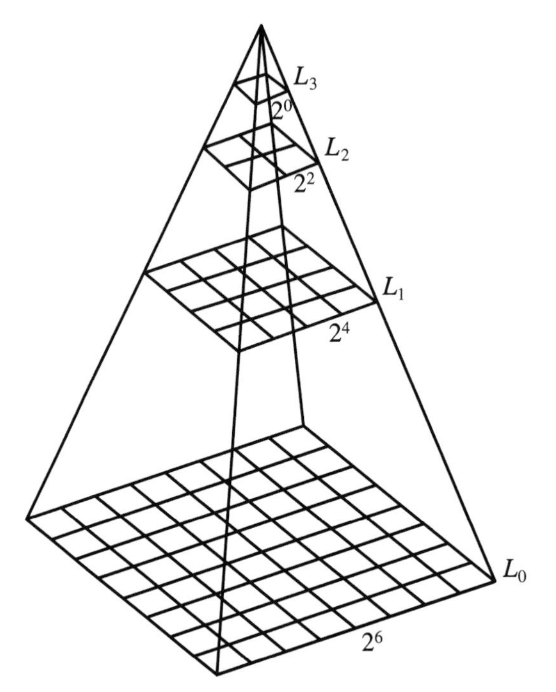

FIGURE 12.14 Pyramid data structure.

a value of 1 is included in four pixels on the lower level, the value of the pixel on the higher level becomes 1.

12.4.2.2 Application of the Pyramid Data Structure

Using a pyramid data structure, image analysis or filtering can be performed. Through the analysis of the higher level, abstract processing is performed. If needed, detailed processing can be performed by using the lower level of the pyramid data. Pyramid data on the higher level is used to achieve a shorter processing time and smaller memory space.

12.4.2.3 Image Analysis Using a Tree Structure

Coding technology can be represented by a tree with four values, i.e., a binary tree. A binary image with a $2^n \times 2^n$ area coded into a tree representation with four values is shown in Figure 12.15.

12.4.2.4 Coding Technique

The coding procedure is performed as follows.

(1) At the root of the tree, the value "1" is output and processed recursively from left to right.

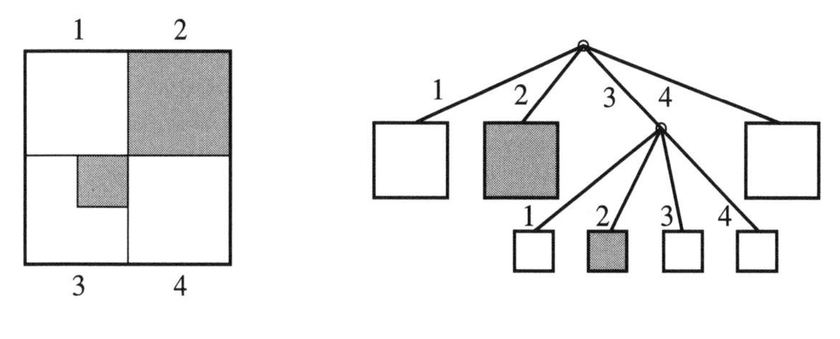

FIGURE 12.15 Coding.

(2) When a node with no leaf is reached, the value "1" is output and the process proceeds to the next step.
(3) When a leaf is reached, the value "0" and a color code (black or white) are output and the process goes on to the next step. When a leaf is a pixel, only a color code is output.

In Figure 12.15, the following coded information is created:

10W0B1WBWW0W

12.4.3 Image Recognition

Computer vision provides the functions for recognizing and identifying an image as a specific object, such as a house, a human, or a road. Human beings recognize an object using their knowledge of the object (e.g., a house, a human, or a road). It is necessary to develop image-recognition technology that can recognize all kinds of objects.

12.4.3.1 Course of Pattern Recognition

The course of pattern recognition is shown in Figure 12.16. After input of an image, noise is eliminated by smoothing, and the size of the image is normalized by using a pyramid data structure. The features of the image are obtained by means of filtering or Fourier transformation. A pattern that matches a standard form is chosen. The most suitable pattern is selected by using dynamic programming and the distance between the input image and the standard image.

12.4.3.2 Pattern Matching

Calculating the Distance Between an Input Image and a Standard Image: We calculate the distances between input image feature points and those of a standard image, or the distances between the input-image's Fourier transform coefficients and those of the standard image. For example, given

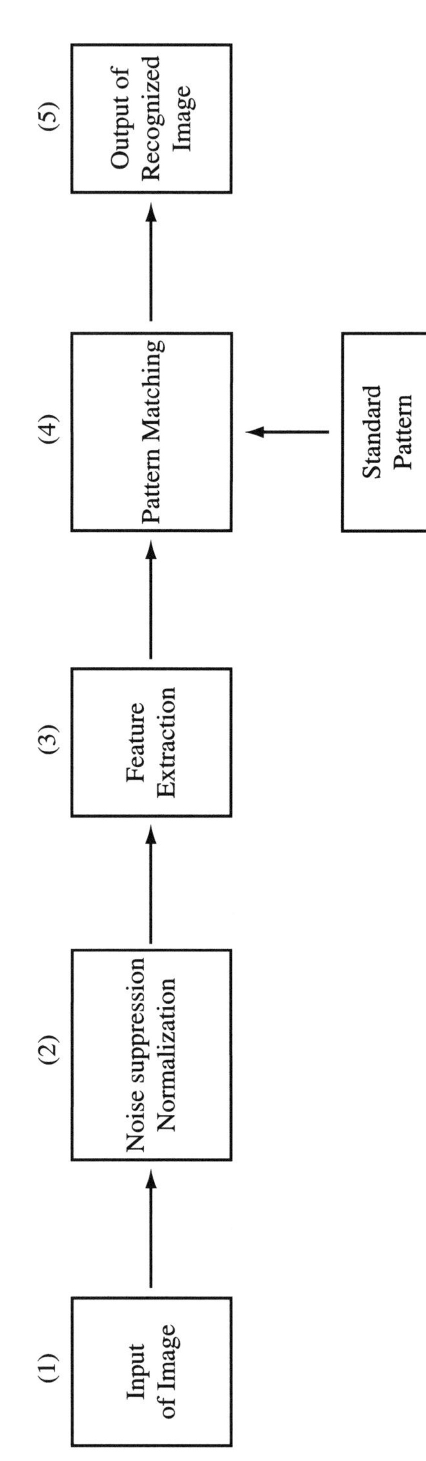

FIGURE 12.16 Pattern recognition.

a coefficient a_I of the Fourier transform of an image and a coefficient s_i of that of a standard image, the distance is determined as follows:

$$d = \sum_{i=1}^{n} | a_i - s_i | \tag{47}$$

Method of Dynamic Programming: In this method, an input image is compared to a standard image. One-to-one correspondences are performed between the pattern of the image and that of a standard image, in order, one by one, as follows:

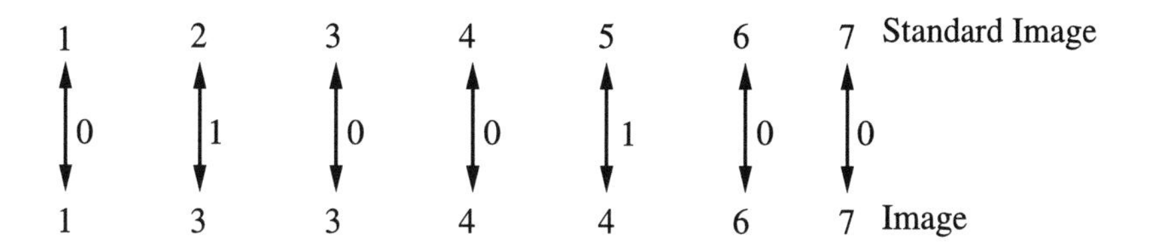

One-to-one correspondences are performed dynamically to get the shortest distance, as follows:

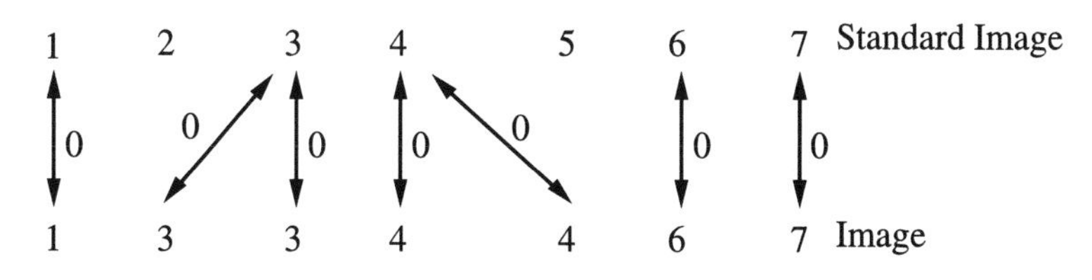

The following correspondences are prohibited:
(1) Corresponding items are crossed:

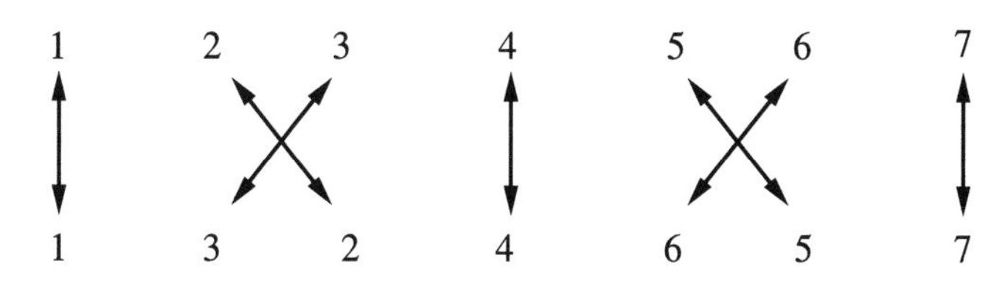

(2) The distance between corresponding items is too large:

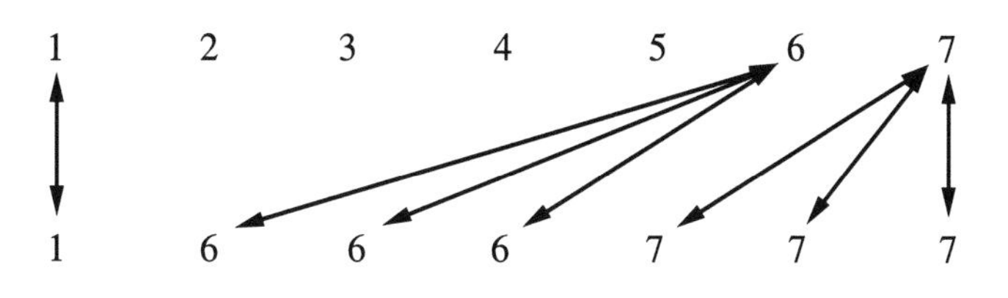

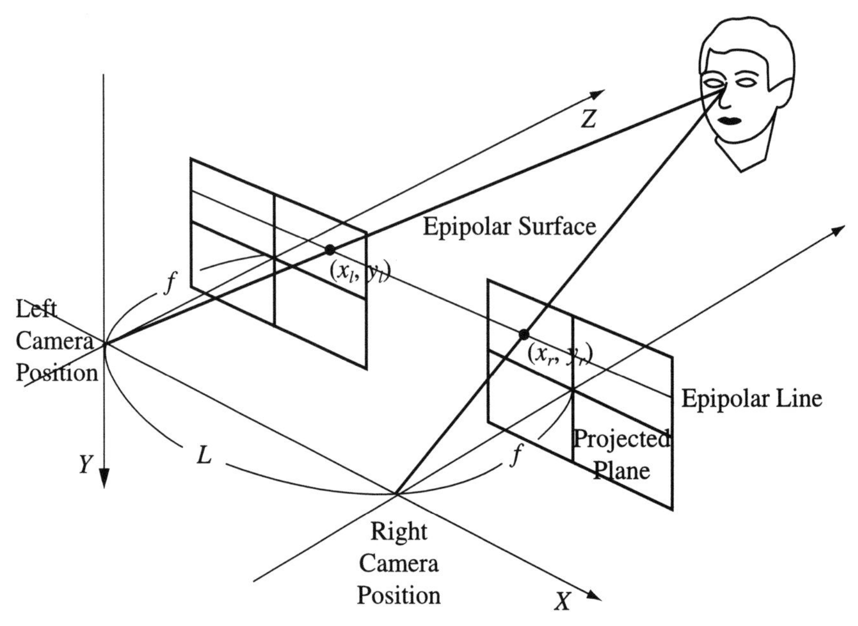

FIGURE 12.17 Stereo analysis.

12.4.4 Stereo Analysis

A line connecting the viewpoint of a camera on the right and that of a camera on the left is called a *baseline* (see Figure 12.17). The distance between the viewpoints is the *baseline length.* A plane connecting three points, such as the viewpoints of two cameras and a point of an object, is an *epipolar surface*. An intersection between an epipolar surface and a projection surface is an *epipolar line*. The point of intersection between a projection surface from the left-hand camera and a line connecting the viewpoint of the left-hand camera with the point of an object (X, Y, Z) is (x_l, y_l). The point of intersection between a projection surface from the right-hand camera and a line connecting the viewpoint of the right-hand camera and the point of the object (X, Y, Z) is (x_r, y_r). When the optical axes of the two cameras are parallel, the following equation holds:

$$\left.\begin{aligned} \begin{bmatrix} xl \\ yl \end{bmatrix} &= \frac{f}{Z}\begin{bmatrix} X \\ Y \end{bmatrix} \\ \begin{bmatrix} xr \\ yr \end{bmatrix} &= \frac{f}{Z}\begin{bmatrix} X-L \\ Y \end{bmatrix} \end{aligned}\right\} \tag{48}$$

Then the following equation holds:

$$\begin{bmatrix} X \\ Y \\ Z \end{bmatrix} = \frac{L}{(xl - xr)} \begin{bmatrix} xl \\ yl \\ f \end{bmatrix} \tag{49}$$

12.4.5 Distance Measurement

The correspondence between the left-hand image and the right-hand image is a big problem in stereo analysis. When a spot beam is projected onto an object and a unique pattern is created on the surface of the object, it is unnecessary to get correspondence between the two images. A spot beam is projected onto the object and the reflection observed by camera. The reflected light is the brightest one seen by the camera. The angle θ between the spot beam and the reflected light is determined. Using the distance between the spot beam source and the camera and the angle θ, the distance between the object and the beam source is calculated by triangulation. This *measurement by spot beam* is shown in Figure 12.18.

Measurement by range finder uses a cylindrical lense to project a spot beam onto the surface of an object. The spot beam cuts the surface of the object and then the shape of the object is determined. The cutting line along the object is determined at the same time, so the required time for measurement is shorter. This method is shown in Figure 12.19.

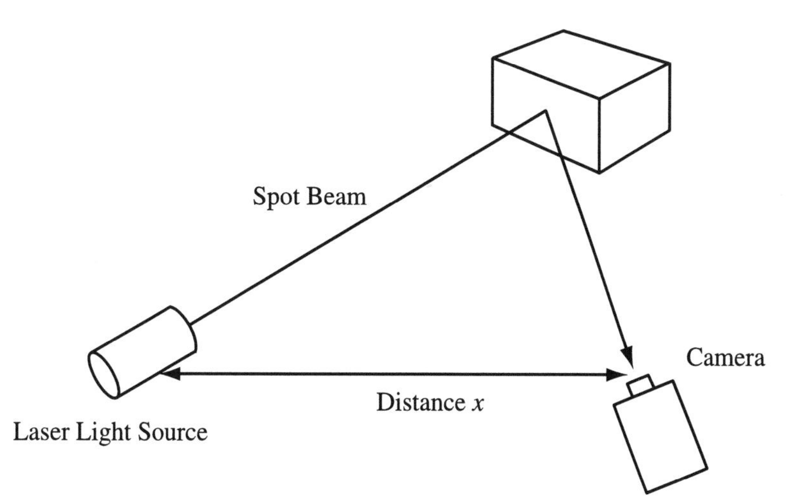

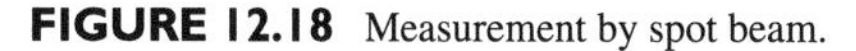
FIGURE 12.18 Measurement by spot beam.

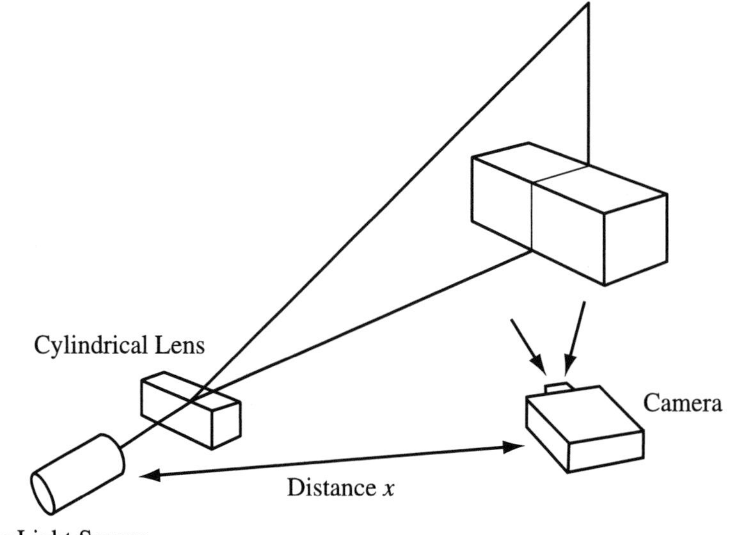

FIGURE 12.19 Measurement by ranger finder.

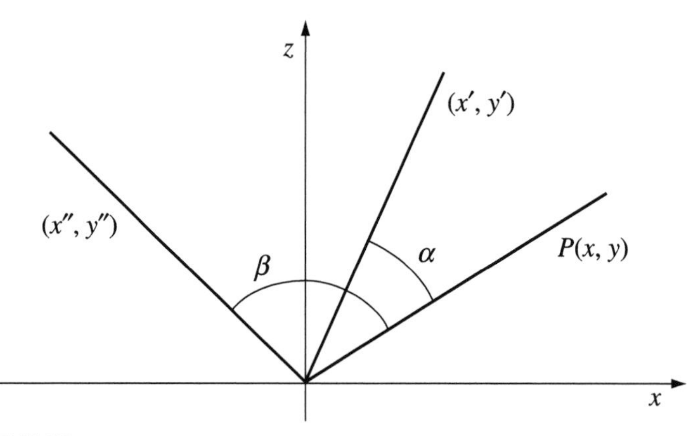

FIGURE 12.20 Object recognition from a 2D image.

12.4.6 Object Recognition

12.4.6.1 Object Recognition from a Two-Dimensional Image

By rotating an object by an angle α, the position x' of P along the x axis is $x' = x \cos \alpha + z \sin \alpha$. Similarly, by rotating an object by an angle β, the position x'' of P along the x axis is $x'' = x \cos \beta + z \sin \beta$ (see Figure 12.20). Combining the two equations, we get, $x'' = ax + bx'$ (50) where $a = [\sin (\alpha - \beta)]/\sin \alpha$ and $b = \sin \beta/\sin \alpha$.

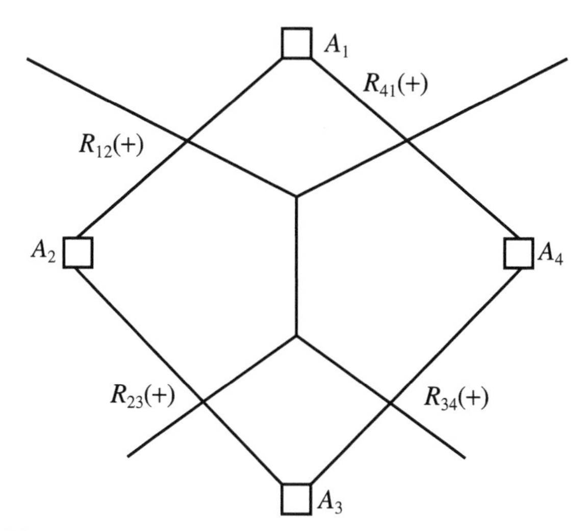

FIGURE 12.21 3D Image description.

Given any a and b for the corresponding points x and x' of two aspects of an image, a corresponding point x'' of a new aspect of the image is obtained via Eq. (50).

$$x'' = ax + bx' \tag{50}$$

12.4.6.2 Object Recognition from a Three-Dimensional Image

Three-dimensional image description is shown in Figure 12.21. The features of a surface, the size of the surface, and the shape of the object are represented as A_1, $A_2, \ldots, A_n$. The relation between two surfaces is notated as R_{ij}. In Figure 12.21, Surfaces A_2 and A_3 are neighboring ones. The relation between A_2 and A_3 is R_{23}, and the shape of A_2 and A_3 is convex.

12.4.6.3 Detection of a Contour Line or a Road From a Map

Both edges of a line or a figure are determined by tracing both edges. Both edges are moved parallel to the figure. In Figure 12.22, curves a and b are determined. The contour line where the distance between a and b is constant is a line. This method enables a contour line or a road to be determined.

12.4.7 Motion Picture Analysis

12.4.7.1 Analysis of Optical Flow

Optical flow: Motion can be determined from examining multiple images. The direction of movement and velocity produce optical flow (see Figure 12.23). Through the detection of optical flow, the movement of an object is determined. Straight optical flow means that an object moves in a straight line [Figure 12.24(a)]. Figure 12.24(b) shows an object rotating.

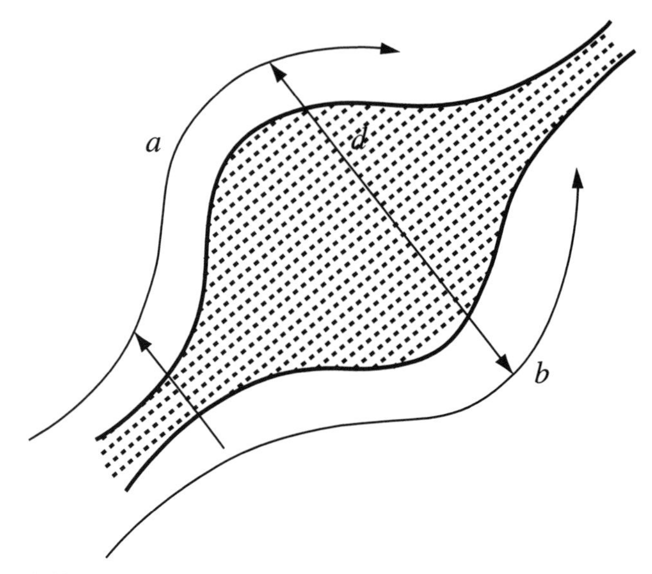

FIGURE 12.22 Tracing method.

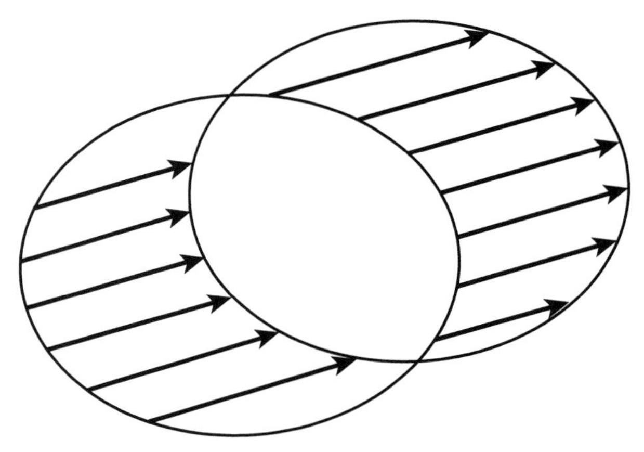

FIGURE 12.23 Optical flow.

Detection of optical flow: Detecting the movement of feature points of two images at different times, allows optical flow to be determined (Figure 12.25).

12.4.7.2 Motion Capture

To recognize human motion, body movement, hand movement, or a facial expression is *motion capture*. Conventionally, markers are put on the face or sensors are put on the body and to register motion. Research on detecting facial expression without the need to wear sensors or markers is also being carried out.

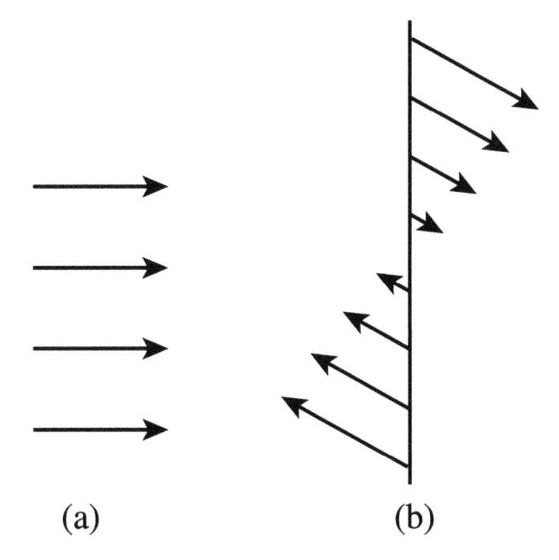

FIGURE 12.24 Optical flow (a) straight-line motion and (b) rotation.

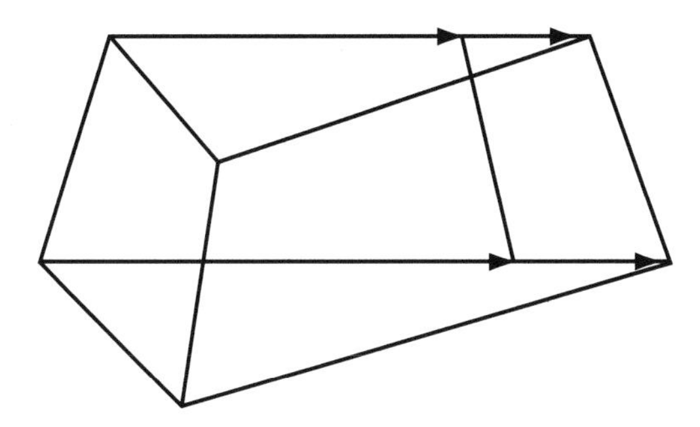

FIGURE 12.25 Optical flow of a 2D image.

Facial expression detection can be done by using Discrete Cosine Transform *(DCT) conversion.* DCT conversion is performed to a rectangular area that includes an eye. When the eye closes, distance along the vertical axis increases. When the mouth opens, distance along the horizontal axis increases. A schematic of facial expression detection is shown in Figure 12.26.

12.5 APPLICATION OF TELESENSATION

12.5.1 Recognition and Synthesis of a 3D Human Image

To provide a virtual space teleconferencing environment or a virtual space distance education environment with realistic sensations, research on how a viewer can get different perspectives on an object or on a human is being carried out. The position of the eye and the direction of view are usually detected by having the participant wear special glasses with sensors. The sensors detect the eye position. The shape of the object is adjusted and the color mapped and displayed on screen to

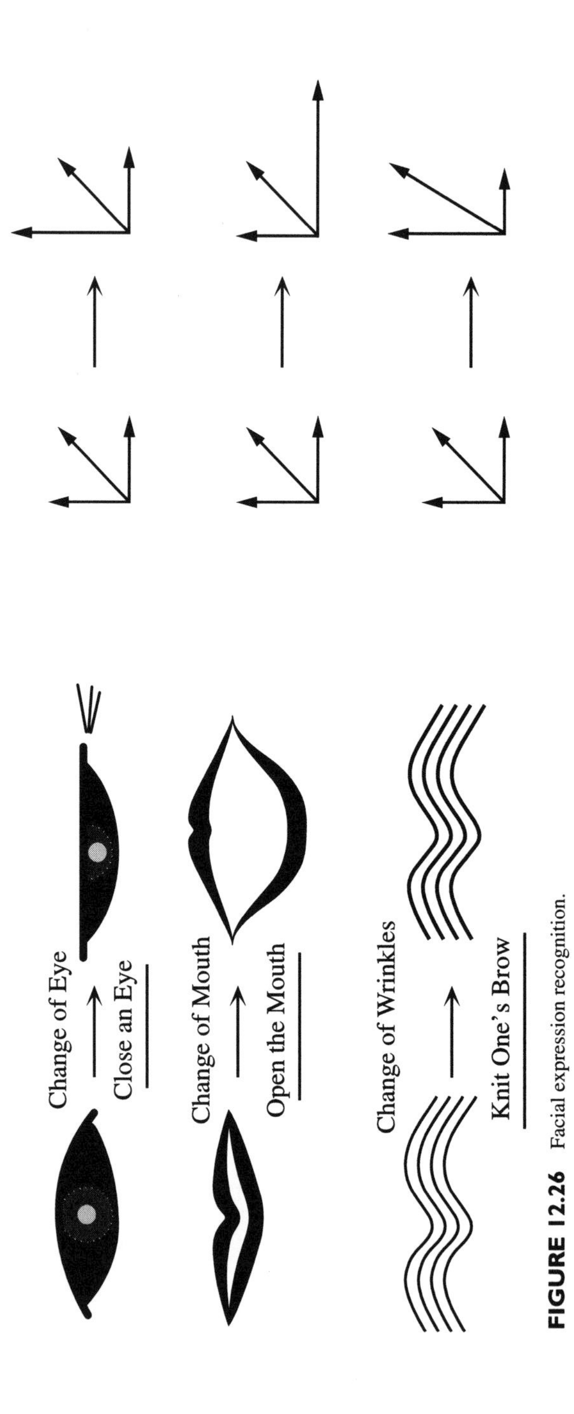

FIGURE 12.26 Facial expression recognition.

match the viewer's perspective. To do this, the shape or color of the object is acquired by computer vision. When the shape of the object is detected, a wireframe model of the object is created and stored in a computer.

Research on synthesis of the human face in real time with realistic sensations is being carried out. To reproduce a human face and its texture in real time using a wireframe model, facial expressions need to be detected. To accomplish this, markers are put around the mouth and eyes. The movement of the markers is detected and facial expressions are determined. Then the wireframe model of the face is adjusted and texture mapped and displayed on screen stereoscopically.

12.5.2 Gesture Recognition

Hand gestures are recognized by wearing a data glove with a sensor. Hand shape is also detected by the data glove. Hand position is detected by a sensor attached to the glove.

Research on gesture recognition without any kinds of devices is being carried out. For example, using two cameras, the left image of a hand shape and the right image are taken. Using the hand shape model, pattern matching between the images taken by camera and the model is performed and a hand shape detected. Two cameras are used to avoid obstruction of the hand shape.

13

CONCLUDING REMARKS

This book has described the idea of the intelligent communication system and its theories and applications. The intelligent communication system represents a fusion of intelligent processing and communication technology. It will provide human-friendly communication services and advanced description methods. At the same time, it will provide a human-friendly communication environment where human beings, real or virtual, are brought together through communication networks for cooperative work. It will play an important role in this century not only in industry but also in our daily lives.

Let us take a look at just how the intelligent communication system will affect industry and our daily lives.

13.1 THE AGE OF THE FIVE SENSES

In 1994, Hakuho-Do of Japan interviewed people to find out how sensitive were their five senses: vision, hearing, taste, smell, and touch. They obtained the following results.

(1) More than 50% thought they were sensitive.
(2) The more experience people had, the more sensitivities they had.

(3) More than 30% thought they wanted to enhance their sensitivities.
(4) People with keen senses were apt to have strong feelings about merchandise when interested in it.
(5) People between the ages of 20 and 30 had keen vision and hearing.
(6) People between the ages of 40 and 60 had keen taste, smell, and touch.
(7) People who were keen in all five senses were apt to purchase merchandise that left a strong impression.

People between the ages of 20 and 30 were born in the 1970s to the 1980s, a substantially advanced period in the economy during which most households purchased television sets and children watched TV. Thus these people were trained by watching and listening to TV. In the 1990s, they played video games. Having spent most of their younger years inside their homes, they had little experience playing games outdoors and did not develop their sense of touch much.

On the other hand, people between the ages of 40 and 60 had little experience watching TV. Instead, they went outside and played baseball in the playground with their friends. And those people, having lived longer, have had richer experiences and also have had the chance to train their senses of touch and taste.

To reiterate, then: people in the age range of 20–30 have keen sight and hearing, whereas, people in the age range of 40–60 have keen taste, smell, and touch. As people between 20 and 30 get older, their senses of taste and smell will be trained.

In the future, people will prefer merchandise that appeals to all five senses, so merchandise that appeals to the five senses will be attractive to them. Hakuho-Do analyzed how people purchased merchandise, with the following results.

(1) In the 1960s and 1970s, people chose merchandise by considering its efficiency, function, and price. If the efficiency or function was reasonable, they purchased it. Good merchandise sold out, whereas bad merchandise did not. Manufacturers produced merchandise that most customers preferred. Manufacturers produced a large amount of popular goods.

(2) In the 1980s, customers purchased their own favorite goods, which were not produced massively. Therefore a variety of goods were developed for sale.

(3) In the 21st century, customers will prefer merchandise that appeals to all of their five senses, and most people will be keen in all five senses. Therefore, the company that produces merchandise that appeals to all five senses will be a winner. In other words, merchandise that appeals to the five senses will be sold and get a majority of the market.

Merchandise that appeals to the five senses will be goods that provide realistic sensations, such as those produced by applying HyperReality technology, and a lot of merchandise will be produced for sale via HyperReality.

Just how will HyperReality develop? In its current stage, a virtual world is created on the screen of a personal computer and a virtual object displayed. A viewer wearing shutter glasses and a data glove gets a stereoscopic view of the object and handles it by hand gesture. In its second stage, a virtual world will be created on the screen of a flat display and the viewer will handle virtual objects by hand gesture. The third stage will be full-screen HyperReality. A wall-size screen will be hung on a wall. HyperReality will become the environment of the room. Through the screen, real people and virtual people will be able to interact, talk together, and conduct cooperative work. In the final stage, a full envelope of HyperReality will be developed. By wearing an "envelope," a person will be able to enter a HyperWorld and communicate for work or play. The envelope will be a kind of data suit with a wearable computer, a data glove, glasses, and wireless communication equipment.

Many kinds of applications will be developed for HyperReality; including HyperClinics, HyperArt museums, and HyperManufacturing.

13.2 THE AGE OF PERSONALIZATION

What might happen if the intelligent communication system is realized and put to practical use? People at different locations could be brought together via the Internet to conduct cooperative work as if they were gathered at the same place. For example, through HyperReality, dress designers at different places around the world could be brought together over the Internet in a HyperDesign room to brainstorm about designing a new dress and to share information about a customer's opinions. At the same time, customers at different places could be brought together via the Internet in a HyperMeeting place to talk about a new dress they would like to purchase. Then designers and customers could be brought together in a common HyperRoom to exchange information about the dress, allowing designers to learn about the dress that customers want to purchase. The designers could then alter the shape or style of a dress or its color or texture and show the revised garment to the customers. Once customers were satisfied with the dress, the design could be sent to a dressmaker to be made. Many kinds of design work could be accomplished in this way, allowing personalized manufacturing to be performed.

How would this work actually be done? In the HyperDesign room and the HyperMeeting room, workstations with display, keyboard, mouse, data glove, and shutter glasses would be installed, enabling designers and customers to communicate. In front of the work station display a virtual world would be generated in which a virtual dress would be displayed. Designers and customers would be able to grasp the dress by hand gesture or mouse and look at it from various angles. According to customers' information, designers could change the shape or color of the dress. Then customers would look at it and give their opinions. A customer would even be able to wear a dress by using his or her image.

Currently customers can go shopping in a store to purchase a ready-made dress. The customers generally can't order a dress fit to them personally. However, through the introduction of HyperReality, they would be able to order directly a personalized dress to be made by a dressmaker.

This should have a great impact on the paradigm of manufacturing. Currently, manufacturing involves mass producing stereotyped products (such as dresses, cars, furniture, and personal computers). Customers then purchase these ready-made products. But in the future, this paradigm would be substantially changed. Customers would ask a manufacturer to produce a personalized product by means of HyperReality and then purchase it from the manufacturer directly. In this way a manufacturer-based service would be changed to a customer-based service.

In addition, customers or subscribers would be able to choose the services they want. For example, if you choose to have caller ID service, you could check what number is calling before you pick up the receiver. Or as a TV subscriber, you could choose what to watch from the many programs broadcast by your broadcasting service company. You could order a dress from a manufacturer directly instead of choosing one from a show window. Thus customers, subscribers, and telephone users would be able to choose their own goods, programs, and callers.

13.3 IMPACT OF THE INTELLIGENT COMMUNICATION SYSTEM ON INDUSTRY

Industrialization among advanced countries has squandered a huge amount of energy, e.g., petroleum, leading to such serious problems as the greenhouse effect. We need to shift to a sustainable economy that does not use so much energy.

The intelligent communication system will be an important tool for achieving a sustainable economy. How it will contribute to this is described in this section.

13.3.1 Presenting Cultural Heritage by HyperReality to the World

Let's take as an example a system for presenting Japanese cultural heritage. Scenes of Japanese cultural heritage, such as Japanese temples, shrines, or the statues of Buddha in Kyoto, will be shot by camera and transmitted to HyperReality centers via the network. At each center, the scenes will be displayed stereoscopically via HyperReality. A viewer will be able to enter the scene and walk through it as if he or she were in Kyoto, giving the viewer the experience of visiting Kyoto virtually by means of HyperReality. Thus people in any country could access any scene of cultural heritage virtually by means of HyperReality, gaining an understanding of the cultural heritage of another country by visiting there virtually. This will make long airplane journeys to Japan unnecessary, saving not only on personal wear and tear but also on the consumption of energy.

To implement such a cultural heritage presentation system, HyperReality centers would be constructed throughout the world and interconnected by the communication network. In each center, tools such as workstations, communication devices, large screens, printers, facsimile machines, telephones, and audio equipment and software for HyperReality would be installed. Then at each center, the cultural heritage information transmitted from, say, Kyoto would be processed, if necessary, and displayed stereoscopically on screen. Visitors would be able to enter the scene and walk through it as if in Kyoto. They could, for example, go to a virtual Kiyomizu-dera temple and enjoy the panoramic view of the mountains surrounding Kyoto. They could visit Touji temple to view the many statues of Buddha that are among Kyoto's national treasures. Through HyperReality, a scene of cultural heritage could be transmitted over the network to a remote place and displayed stereoscopically on screen. Viewers could send questions about the Kyoto temple or the statues of Buddha to the Kyoto center and interactively get answers from Kyoto over the network. They could access any center via the network and ask that center to send any scene to their terminal. A person who wanted to visit, say, Kyoto could reserve a flight and a hotel room from his or her terminal.

13.3.2 Developing Products for an Aging Society

This century will see a move to the aging society, especially in advanced countries. For example, in Japan, people over 65 years old will exceed 25% of the total population in 2020. Some of them will surely want to continue working after age 65, so it will be important for them to have job opportunities. Others will prefer to enjoy a life of retirement, making new friends by exchanging information and enjoying hobbies. Some will be disabled and need to be taken care of. How will the intelligent communication system contribute to fulfilling these needs of an aging society? As to how to provide job opportunities to old people, some of them would like to work at home. To achieve this, a human-friendly working environment could be developed by using HyperReality. A workstation that provides a human-friendly work environment could be installed in the home. On the work station screen a virtual space could be created. By wearing a data glove, the viewer could handle a virtual object by hand gesture and gain a stereoscopic view of the object by wearing special glasses. The user could interact with the computer by means of natural language in a human-friendly fashion. The workstation could enable virtual meetings with colleagues.

People could interact over the network, playing games or viewing pictures drawn by someone else. People in a common interest group could exchange over the network the images of drawing pictures together and enjoy watching them.

Home care service will be one of the important issues in this century. This will include not only visiting and seeing patients but also monitoring them by means of HyperReality. Doctors or caregivers and patients could be brought together via the network. Care would be given as if everyone were at the same place. A patient's temperature and pressure could be measured and the results sent to the doctor or caregiver. When action is needed, the doctor or caregiver would then visit the patient's house and perform the appropriate action.

In addition to this care system, an entertainment service could be provided to patients by using HyperReality via workstations installed in their home; patients would be able to access a doctor's or caregiver's center to ask for care. While staying at home, the patients could also visit places or do things remotely.

13.3.3 Construction of a Platform for HyperReality

To develop the aforementioned services, a platform is needed to provide all kinds of applications, including design work, elderly or disabled care, and entertainment. For example, in the case of design work it should provide human-friendly design tools not only to designers but also to customers. And human-friendly interfaces should be available to the elderly or the disabled. The platform for design should enable designers and customers to interact and collaborate on the design of new goods that satisfy customers' demands.

Currently, a manufacturer produces goods that are distributed to customers via wholesalers and retailers. A customer goes to a shop to purchase goods. Through the introduction of a HyperReality platform, manufacturing and distribution would be changed substantially. A customer would interact with a manufacturer to ask them to produce a particular item, maybe even cooperating on the design, as if they were in the same place. Once the design is complete, it would be sent on to a factory to be manufactured, after which it would be sent directly to the customer. This would give customers access to any manufacturer in the world, enabling them to ask the manufacturer to produce the goods desired.

By enabling customers to choose one of the best manufacturers throughout the globe, competition among manufacturers should be stimulated. To survive this competition, manufacturers will have to strive to make better goods. They will have to employ the best creative designers to beat their competitors. Manufacturers will require information concerning customers' demands and will be forced to give the best solutions to customers. Manufacturers will have to improve their skills and produce better goods. All of this work will be done via the network, obviating any need to get together physically for the design work.

13.3.4 Development of New Services

HyperReality will enable HyperShopping and HyperTravel to be provided. HyperReality will create a HyperWorld and one or more coaction fields where people, real or virtual, will be brought together via the network to interact and work together as if in the same place. Any place where a HyperRealty-based workstation is installed, people will be able to go to a HyperArt Museum or shop on Fifth Avenue in New York City or the Ginza in Tokyo, all done virtually. To accomplish this, Fifth Avenue will be shot by camera and sent to viewers over the network. At the viewers station, the scene of Fifth Avenue will be displayed stereoscopically on screen. They will be able to enter a virtual store and enjoy walking through and shopping. They will be able to see the merchandise, look at it from any angle, and grasp it by hand gesture. They will be able to order it through the network.

When a Japanese customer visits a New York virtual shop via the network and talks with staff in the shop, they will have to speak English, otherwise they will be unable to communicate with the staff. In that case, it will be desirable to provide a language interpreter to translate between Japanese and English.

Participants will be able to go to Waikiki, for instance, for a virtual swim by means of HyperReality. A disabled person will be able to go to a HyperArt Museum to enjoy the painting, even while staying at home.

13.4 IMPACT OF THE INTELLIGENT COMMUNICATION SYSTEM ON SOCIETY

Our daily lives have been changing rapidly due to the introduction of information technology. The number of dot-com companies is increasing. Book stores, variety stores, general stores, and food stores all open in cyberspace. Hotel or airline flights reservations can easily be booked online. The broadcasting of TV programs is now monitored by computer. The small office/home office has become a typical example of telecommuting. Thus has information technology been applied to a variety of business fields. The intelligent communication system is a more advanced system that combines artificial intelligence and communication technology. How it will change our lives is described in this section.

13.4.1 HyperShopping

A workstation with a large screen will be installed in each customer's house, with the station linked to any shop via the Internet. Customers will open the Web page of a store that sells the goods they want to purchase. They retrieve information about the goods, view the goods in 2D or 3D images or stereoscopically on screen. They will grasp them by hand and look at them from any angle. If they like what they see, they can purchase the items. If a woman wants to buy a dress, she can try it on her virtual image, which is exactly her size.

13.4.2 Getting the Latest Information

By accessing a Web page, people can get the latest information on business, entertainment, shopping, and the like. But through the introduction of the intelligent communication system, the access will be more human friendly. Search engines will be enhanced by the introduction of intelligent technology. Hotel or airline reservations will be made over the Internet via a human-friendly interface.

13.4.3 Small Office/Home Office (SOHO)

HyperReality will change the small office/home office (SOHO) substantially. Through the installation of a HyperReality-based workstation with a large screen, a home will become a small satellite office. The workstation will be linked to the main company for exchange of information between the small office and the company. The employee will be able to have a videoconference with colleagues via the network. A virtual space will be created on screen where one or more virtual objects will be displayed. Wearing a data glove, participants will be able to grasp such an object by hand. They will be able to change its shape or color or texture. If needed, they can do cooperative work over the network as if they were in a same place. They will be able to link to any site anyplace in the world and collaborate with partners around the world. Therefore SOHO will become global.

In the future, the screen will be large, and workstations will have shutter glasses, a data glove with sensors, a mouse, a joy stick, and a keyboard. The workstations will be efficient and less expensive, too. The Internet will provide high-speed transmission and be less expensive. Three-dimensional images of ourselves, of goods, and of scenes of cultural heritage will be transmitted over the Internet in real time and processed in real time by our workstations.

13.4.4 Changing a Hierarchically Structured Company into a Networked Company

Companies are typically highly structured organizations, with individuals organized into sections and the sections organized into departments. Orders from top management are communicated to the departments and then from each department to its sections. Finally the orders are communicated to members in each section. All of this type of communication from top to bottom takes time.

In the information age, information flow will be changed substantially. Anyone can communicate with anybody else over the Internet, enabling an order to be communicated to any member directly from top management. And answers or comments can be sent directly back to top management in a short time. This means that, via the Internet, communication among members of a company is easily and quickly accomplished. The Internet enables interaction and communication among partners spread around the world, enabling all to extend their field of business. This interaction and communication among people enlarges the possibilities for creating new businesses, ideas, or job opportunities.

All of this requires the development of more human-friendly interfaces. The intelligent communication system, because it provides human-friendly interfaces, will play an important role in achieving this.

13.5 MULTIMEDIA-BASED SOCIETY IN THE 21ST CENTURY

Technology advanced greatly in the 20th century. Heavy industry, the steel industry, and the computer industry played major roles in enhancing our lives, as did high-tech devices and the growth of the economy.

All of this involved a huge consumption of energy, especially petroleum. As a result, we have the greenhouse effect: average temperature on earth rose 4°C and sea level rose about 1 meter during the 20th century. If we continue to use petroleum on the same scale, the earth's temperature will rise more than 6° this century. We have to change the style of our activities and find new methods if we're to survive.

One solution is to make information technology the basis of our activities. In the 21st century, information technology will greatly advance. Broadband networks will be constructed and put into practical use at lower cost. At the same time, computers will be improved, and all information, including multimedia information, will be easily and efficiently handled. In every household, a workstation that is efficient and less expensive will be installed. Information will be rapidly transmitted among workstations over the network.

People will be able to get involved in virtual space, going anyplace in the world—virtual museum, a virtual school, a virtual shop, a virtual company. They will be able to view great paintings, study mathematics, purchase apparel, and do collaborative work with partners in foreign countries. They will be able to watch a three-dimensional television program, able to enter the scene and act along with those on the broadcast. They will be able to wear a data suit—a combination of a data glove, a small screen, a computer, a graphic engine, and communication equipment—by which to communicate anywhere with a company, a customer, a store, or a school. This wearable computer and communication equipment will enable them to visit a virtual Waikiki Riviera and enjoy sightseeing or shopping, go to a golf course, or play with friends just as if they were actually there.

Newspapers will change substantially, too. Online newspapers will be delivered over the Internet to every home. Customers will be able to read the paper or display an object in the paper on the screen stereoscopically and not only enjoy watching it but also handle it by hand gesture and look at it from all around.

People will be able to study whenever they want. HyperClasses will be brought to people's homes via the Internet. A teacher and students at different locations will be brought together over the Internet as if they were gathered in the same classroom. It will be unnecessary for people to go abroad to study. They will be able to attend any class, even if conducted in a foreign country. Thus people will be able to go anywhere to attend a class, go shopping, travel, or do business. The experience of traveling abroad will be enhanced by HyperReality.

The intelligent communication system will penetrate all human activities and change our lives. Through HyperReality, human beings will be able to break the barriers of time and space.

The following areas will need to be developed to establish a multimedia-based society.

(1) *Information infrastructure:* Equipment such as ATM switching systems and fiber-optic networks have been constructed to provide broadband services, especially in advanced countries. In order to transmit multimedia information in

real time, high-speed networks and highly efficient switching systems such as ATM switching systems have to be installed everywhere. High-speed local area networks are also needed to provide high-speed multimedia services.

(2) *Database technology:* The amount of data in the database of multimedia information will be huge. Motion pictures, 3D image data, and still pictures require databases that are huge as compared with textual information. It is desirable to develop a multimedia database system with the storage space for a huge amount of information as well as high-speed processing capabilities.

(3) *Image recognition and reconstruction technology:* It is very important to develop image recognition and synthesis technology. Through HyperReality, human beings, real or virtual, will be brought together via the Internet to do cooperative work as if they were gathered in the same place. First the shape and texture of a person's image is measured by image recognition technology. Current image recognition devices include 3D cyberware by which the shape of a 3D object is measured and its color and texture determined. However, cyberware can measure a 3D object whose height is no greater than 30 centimeters. More powerful image recognition devices need to be developed. When a person's shape, color, and texture are determined, the information will be stored at the workstation. During the session, the person's facial expression or body motion will be recognized by means of image recognition technology. Using this information, the 3D image will be adjusted by image synthesis technology and displayed in a coaction field. This is how all objects will be measured by image recognition technology. The shape, color, and texture will be determined. When the motion of the object is detected, the shape will be adjusted by means of image synthesis and the texture mapped and displayed in a coaction field.

(4) *Human—machine interface technology:* In HyperReality, a person will be able to handle a virtual object by hand gesture. Wearing a data glove with sensors, the person's hand motions and finger motions will be detected. Using the information, the image's hands and fingers will be adjusted and displayed on screen. HyperReality will give a more advanced human-friendly interface between human and computer. Natural language processing, speech recognition, and handwritten-text recognition will provide human-friendly interfaces, especially for disabled persons. These kinds of multimodal human–machine interfaces will give more chances for interaction between people and computers.

13.6 BRIDGING THE GAPS BETWEEN THE HAVES AND THE HAVE-NOTS

In the information age anyone should be able to access the Internet and communicate. However, there are people who lack access to the network. Especially in the developing countries, most of these lack such access because they do not have a computer or enough money to access the network.

This problem of the haves vs. the have-nots should be resolved by every conceivable means. It is predicted that the gap will increase between the haves and the have-nots as the information age progresses. To overcome this gap, the following measures should be taken.

(1) The cost of computers and information infrastructures must become affordable so everyone can access the network and communicate and work together over the network.

(2) Human-friendly human–machine interfaces should be developed by which everyone can access the network. To achieve this, an intelligent agent should be installed in the system that will give an understandable introduction on how to gain access to and use the system.

(3) Education to enhance information literacy should be conducted. Every one has his or her skill level in using a computer and the network. Personal education should be planned and conducted based on a person's skill level. In general, young people can easily understand computers and the network because they are accustomed to computers as game machines. In contrast, many older people have difficulty understanding them. Thus, education should match skill level. And because gaps exist among countries, international conferences and forums must be held to ensure that these gaps dissolve in the future.

13.7 LIGHT AND SHADOW OF MULTIMEDIA-BASED SOCIETY

As mentioned many times already, the intelligent communication system will have a great impact on our daily lives and ways of life. Through its introduction, the paradigms around us will be greatly changed. In particular, a manufacturer-based economy will become a customer-based economy. Everyone, young or old, will have access to a computer and the communication network. Everyone will be able to join together to work cooperatively over the network as if we were gathered in the same real place. The barriers of time and space will disappear in the information age. The styles of our daily lives and even our ways of thinking will be changed. The combination of physical reality and virtual reality will bring substantial surprise and excitement to human beings. We will enter this new world and have experiences we've never had before.

On the other hand, many of the problems of industrialization, such as energy shortage and population explosion, have been pointed out and focused on. If we want to survive on earth, we should think about these problems, seek solutions, and improve the current situation. If we try to solve these problems, we will surely find the solutions. But if we continue to act as we have been acting, the pollution, environmental degradation, and excessive energy consumption will be big problems in the future.

We have to change our ways to solve these problems. The intelligent communication system is an important step in the right direction. By shifting to new modes based on the intelligent communication system, we should be able to alleviate such problems as out-of-control energy consumption. This will have a positive effect.

However, it could have a negative effect as well. The fusion of physical reality and virtual reality could make it very difficult to distinguish between real life and virtual life. There is always the possibility that the intelligent communication system will be used for illegal purposes, such as pornography. We must improve our abilities to discriminate between legal and illegal products if we want to survive and enjoy life in the future.

REFERENCES

Adiv, G. (1985). Determining three-dimensional motion and structure from optical flow generated by several moving objects, *IEEE Trans*. Vol. PAMI-7, No. 4, pp. 384–401.

Alty, J. L., and Coombs, M. J. (1984). *Expert Systems: Concepts and Examples.* Manchester: National Computing Centre.

Azuma, R. T. (1997). "A survey of Augmented Reality," *Presence, Teleoperators and Virtual Environments,* 6, 4: 355–85, Massachusetts, MIT Press.

Ballard, D. H. (1981). Generalizing the Hough transform to detect arbitrary shapes, *Pattern Recognition*, Vol. 13, No. 2, pp. 111–122.

Barr, A., and Feigenbaum, E. A. (eds.) (1981). *The Handbook of Artificial Intelligence,* Vol. 1. Stanford: Pitman.

Barr, A., and Feigenbaum, E. A. (eds.) (1982). *The Handbook of Artificial Intelligence,* Vol. 2. Stanford: Pitman.

Bellman, R. (1957). *Dynamic Programming*. Princeton, NJ: Princeton University Press.

Bobrow, D. G., et al. (eds.) (1975). *Representations and Understanding.* Orlando, FL: Academic Press.

Bundy, A., et al. (1978). *Artificial Intelligence: An Introductory Course.* Edinburgh: Edinburgh University Press.

Charniak, E., et al. (1980). *Artificial Intelligence Programming*. Hillsdale, NJ: Erlbaum.

Cohen, P. R., and Feigenbaum, E. A. (eds.) (1982). *The Handbook of Artificial Intelligence,* Vol. 3. Stanford: Pitman.

Davis, J. C., and McCallagh, M. S. (1975). *Display and Analysis of Spatial Data.* New York: Wiley.

Duda, R. O., and Hart, P. E. (1973). *Pattern Classification and Scene Analysis.* New York: Wiley, p. 271.

Fischer, M. A., and Bolles, R. C. (1981). Random sample consensus, *Comm. ACM,* Vol. 24, No. 6, pp. 381–395.

Gilmour, J. (1988). Intelligent Network/2, *IEEE Communications Magazine,* Vol. 26, No. 12.

Golay, M. J. E. (1969). Hexagonal pattern transformation, *IEEE Trans. Computers*, Vol. C-18, No. 8, pp. 733–740.

Gray, R. M. (1984). Vector quantization, *IEEE ASSP Magazine*, Vol 32, pp. 4–29.

Haralick, R. M. (1972). Textual features for image classification, *IEEE Trans. SMC*, Vol. SMC-3, No. 6, pp. 145–168.

Harman, P., and King, D. (1985). *Expert Systems.* New York: Wiley.

Hays-Roth, F., et al. (eds.) (1983). *Building Expert Systems.* Reading, MA: Addison-Wesley.

Horn, B. K. P., and Schunck, B. G. (1981). Determining optical flow, *Artificial Intelligence,* Vol. 17, pp. 185–203.

Jarvis, R. A. (1983). A perspective on range finding techniques for computer vision, *IEEE Trans.*, Vol. PAMI-5, No. 2, pp. 122–139.

Lehshcheller, H., and Franbe, U. (1987). Color picture coding—algorithm optimization and technical realization, *Frequenz,* Vol. 41, no. 11/12, pp. 291–299.

Lindsay, P. H., and Norman, D. A. (1977). *Human Information Processing: Introduction to Psychology,* 2nd ed., Orlando, FL: Academic Press.

Linnainma, S. (1988). New efficient representations of photographic images with restricted number of gray levels, *Proc. 9th ICPR,* p. 143.

McCarthy, J., and Hayes, P. J. (1969). Some philosophical problems from the standpoint of artificial intelligence, *Machine Intelligence,* Vol. 4, pp. 463–502.

McCorduck, P. (1979). Machines Who Think. New York: W. H. Freeman.

Minsky, M., and Papert, S. (1969). *Perceptrons.* Cambridge, MA: MIT Press.

Minsky, M. (1975). A framework for representing knowledge, *The Psychology of Computer Vision,* pp. 211–277. New York: McGraw-Hill.

Milgram, P., and Kishino, F. (1994). "A Taxonomy of Mixed Reality Visual Displays," *IEICE Transaction on Information Systems,* E77-D, 12, Tokyo, IEICE Japan.

Nilsson, N. J. (1980). *Principles of Artificial Intelligence*. Paolo Alt.: Toiga.

Pratt, W. K. (1991). *Digital Image Processing*, pp. 149–155, New York: Wiley.

Rich, E. (1983). *Artificial Intelligence.* New York: McGraw-Hill.

Riecken, D. (ed.) (1994). Intelligent agents, *Communications of the ACM,* Vol. 37, No. 7, pp. 18–106.

Sakoe, H. (1979). Two-level DP matching, *IEEE Trans. Acoust. Speech and Signal Process*, Vol. ASSP-27, pp. 588–595.

Schank, R. C. (1975). *Conceptual Information Processing,* Amsterdam: North-Holland.

Shapiro, S. C., et al. (eds.) (1991). *Encyclopedia of Artificial Intelligence,* 2nd ed., New York: Wiley International.

Simons, G. L. (1984). *Introducing Artificial Intelligence*. Manchester: National Computing Centre.

Terashima, N. (1993). 'Telesensation—a new concept for future telecommunications,' *Proc. TAO First International Conference on 3D Image and Communication Technologies*. Tokyo, Telecommunication Advancement Organization.

Terashima, N. (1994). Virtual space teleconferencing system, *Proc. 3rd Int. Conf. Broadband Islands*. Amsterdam: North-Holland.

Terashima, N. (1994). Virtual space teleconferencing system—A distributed virtual environment, *Proc. IFIP Congress 94*. Amsterdam: North-Holland.

Terashima, N. (1995). HyperReality, Proc. Int. Conf. Recent Advances in Mechatronics, New York: Springer-Verlag.

Terashima, N. (1995a). "HyperReality," *Proc. International Conference on Recent Advances in Mechatronics.* Istanbul, School of Computing and Technology Publications.

Terashima, N. (1995b). 'Telesensation—fusion of multimedia information and information highways,' *Proc. International Conference on Information Systems and Management of Data.* Heidelberg; Springer.

Terashima, N., Ohya, J., Kitamura, Y., Takemura, H., Ishii, H., Kishino, F. (1995). "Virtual space teleconferencing: real time representation of 3D human images," *Journal of Virtual Communication and Image Representation*, 6, 1:1.

Terashima, N., and Altman, E. (eds.) (1996). *Advanced IT Tools*. New York: Chapman & Hall.

Terashima, N. (1998). HyperClass—An advanced distance education platform, *Proc. IFIP Teleteaching 98*.

Terashima, N. (1999). An experiment of virtual space distance education system using the objects of cultural heritage, Piscataway: *IEEE Computer Society Proc. IEEE Int. Conf. Multimedia Systems 99*.

Terashima, N. (1999). An experiment of virtual space distance learning system, *Proc. Annual Conf. Pacific Telecommunication Council 99*. Honolulu: PTC.

Tiffin, J., and Rajasingham, L. (1995). *In Search of Virtual Class*. London: Routledge.

Winograd, T. (1972). *Understanding Natural Language*. Orlando, FL: Academic Press.

Winston, P. H. (ed.) (1975). *The Psychology of Computer Vision*. New York: McGraw-Hill.

Winston, P. H. (1984). *Artificial Intelligence,* 2nd ed., Reading, MA: Addison-Wesley.

Winston, P. H. (1992). *Artificial Intelligence,* 3rd ed., Reading, MA: Addison-Wesley.

Witkin, A. P. (1981). Recovering surface shape and orientation from texture, *Artificial Intelligence*, Vol. 17, pp. 17–45.

INDEX